Praise for *The Greenhouse and Hoophouse Grower's Handbook*

"Hats off to Andrew Mefferd and his comprehensive guide to growing in a protected environment! While covering all the basics, this book takes growers to the next level, with in-depth discussions on the physical environment, plant biology, and their intriguing relationship. Once hard-to-find info on topics such as plant steering and grafting is presented thoroughly and clearly. Importantly, the last chapters cover the practical ins and outs of growing eight of the most significant crops for protected culture. Every grower will learn lots from this book; I sure did."

— RICHARD WISWALL, author of
The Organic Farmer's Business Handbook

"Andrew Mefferd's book fills a gaping void in the literature for market growers. I highly recommend it to anyone growing in greenhouses, or who aspires to. With experience few others have, Mefferd explains growing techniques used in advanced greenhouses, and then shows how smaller-scale growers might put them to use. I kept a pencil by my side and plan to use lots of ideas on our own farm."

— BEN HARTMAN, author of *The Lean Farm*

"This is an important book that will give growers the tools and resources to increase production and profits in protected-culture environments. In one book, Andrew has packaged the detailed, technical growing information that took us years and thousands of dollars to acquire for our farm. This book will be one I refer to and recommend often."

— MICHAEL KILPATRICK, farmer; cofounder,
In the Field Consultants

"*The Greenhouse and Hoophouse Grower's Handbook* is a must-read for beginners as well as a valuable, up-to-date resource for experienced growers. Mefferd has written a comprehensive overview of the elements of growing crops under cover. So much of the food available now is from greenhouse production, and most from faraway lands. This book will help you be the one growing crops for your local market."

— ERIC SIDEMAN, PhD, crop specialist,
Maine Organic Farmers and Gardeners Association

"Although greenhouse and high-tunnel food production is a well-established industry, small and beginning growers have never had easy access to information about how the professionals do it. Andrew Mefferd has bridged that gap with this important new book. He makes technical information accessible in a lively and lucid style. Everyone who owns or plans to buy a greenhouse or hoophouse should read this book."

— LYNN BYCZYNSKI, author of *Market Farming Success*
and *The Flower Farmer*; founder, *Growing for Market*

"With this book, Andrew Mefferd, a firm supporter of the local food movement, offers a knowledgeable contribution to help growers (especially in cold climates) develop skills to deal with challenges and reap the benefits of protected cropping. This is not a generalist hoophouse book but a menu where growers can select specialized professional practices to suit their situation, whether growing microgreens; grafting, pruning, and trellising vine crops; or 'crop steering' to select for leaf growth or fruit development."

— PAM DAWLING, author of *Sustainable Market Farming*; contributing editor, *Growing for Market*

"The production of organic vegetables using protected culture to modify the natural environment and optimize plant growth is one of the most highly productive systems for organic vegetables. This handbook provides a broad spectrum of knowledge on growing structures and climate control as well as propagation, pruning, trellising, crop steering for maximum production, and grafting plants for natural disease control. For aspiring market gardeners, young and old, *The Greenhouse and Hoophouse Grower's Handbook* is a must-read!"

— DR. MERLE H. JENSEN, professor, Controlled Environment Agriculture Center, The University of Arizona

"Finally, a seasoned greenhouse grower has taken the time to share professional greenhouse techniques with the small-scale farming community. Such valuable work was a long time in the coming! Whether you're currently growing in a greenhouse or hoophouse or planning to do so (and you should!), this book will teach you the best practices. It concisely and methodologically demystifies the all-encompassing skill set that you need to become successful at growing lucrative crops in protected space. Andrew Mefferd knows his stuff, and his book is absolute gold. Can't recommend it enough."

— JEAN-MARTIN FORTIER, author of *The Market Gardener*

"Finally! A book that makes the specialized and highly refined techniques of the big European and North American hothouse growers accessible to farmers, market gardeners, and growers at all scales, with clear explanations of the practices and why they work. It took me twenty years to figure out half of this on my own; I'm glad I don't have to wait another twenty to figure out the rest."

— JOSH VOLK, author of *Compact Farms*; consultant, Slow Hand Farm

"When we started our wholesale greenhouse back in 1974, the best references were the equipment catalogs, and we pored over them each night. Now Andrew Mefferd has created the most complete greenhouse grower's manual in the world. It contains the full spectrum of proper greenhouse growing for organic vegetable production and all its factors and, importantly, is detailed from a foundation of experience. This book is a fact-filled treasure, accessible for all. Thank you for debriefing, Andrew. You are a hero of horticulture!"

— ALICE DOYLE, cofounder, Log House Plants

The Greenhouse *and* Hoophouse Grower's Handbook

The Greenhouse *and* Hoophouse Grower's Handbook

Organic Vegetable Production Using Protected Culture

ANDREW MEFFERD

Chelsea Green Publishing
White River Junction, Vermont

Project Manager: Alexander Bullett
Developmental Editor: Michael Metivier
Copy Editor: Laura Jorstad
Proofreader: Rachel Shields
Indexer: Linda Hallinger
Designer: Melissa Jacobson
Page Composition: Abrah Griggs

Printed in the United States of America.
First printing February, 2017.
10 9 8 7 6 5 4 20 21 22

Our Commitment to Green Publishing
Chelsea Green sees publishing as a tool for cultural change and ecological stewardship. We strive to align our book manufacturing practices with our editorial mission and to reduce the impact of our business enterprise in the environment. We print our books and catalogs on chlorine-free recycled paper, using vegetable-based inks whenever possible. This book may cost slightly more because it was printed on paper that contains recycled fiber, and we hope you'll agree that it's worth it. *The Greenhouse and Hoophouse Grower's Handbook* was printed on paper supplied by LSC Communications that contains at least 10% postconsumer recycled fiber.

Library of Congress Cataloging-in-Publication Data
Names: Mefferd, Andrew, author.
Title: The greenhouse and hoophouse grower's handbook : organic vegetable production using protected culture / Andrew Mefferd.
Description: White River Junction, Vermont : Chelsea Green Publishing, [2017]
Identifiers: LCCN 2016040407 | ISBN 9781603586375 (pbk.) | ISBN 9781603586382 (ebook)
Subjects: LCSH: Greenhouse gardening. | Vegetable gardening.
Classification: LCC SB415 .M375 2017 | DDC 635.9/823--dc23
LC record available at https://lccn.loc.gov/2016040407

Chelsea Green Publishing
85 North Main Street, Suite 120
White River Junction, VT 05001
(802) 295-6300
www.chelseagreen.com

To my parents, with love,
for reading to me from the very beginning

Contents

Acknowledgments

This book is the product of many experiences over the last fifteen years or so. Without the contributions of any of these people, it would not have been possible.

First I would like to thank my wife and farming partner, Ann. As our farm was the test kitchen for much of what is in this book, it wouldn't have happened without her love and support in and out of the greenhouse. Thanks also for slogging through early versions of the manuscript, and making many excellent suggestions along the way.

Satyam Azad of Azad Photo made the images of the grafting sequences look great. Though he usually is a photographer of jewelry and art, it turns out he works on the right scale for seedlings, too. Look him up at www.azadphoto.com.

I'd also like to thank Rob Johnston and all my former colleagues at Johnny's Selected Seeds. The seven years I spent working at the research farm were a time of great learning, which provided much of the understanding that informs this book.

Founder and former editor of *Growing for Market* magazine Lynn Byczynski encouraged me to submit this manuscript, and suggested Chelsea Green as the publisher. For that and much other guidance I would like to thank her.

Michael Metivier, Ben Watson, and the whole team at Chelsea Green deserve a great deal of thanks. Now I understand why writers thank their editors. Their hard work made this a much better book.

If this book leaves you with questions, the University of Arizona Controlled Environment Agriculture Center (CEAC) is a good place to look for answers. I must thank Merle Jensen and the entire staff of the CEAC for being so generous with their time and knowledge when I had questions.

Last but not least, I want to thank the many growers who took time out of their busy days to share insights from their body of knowledge. No one understands plants like someone who has to please them for a living.

Introduction

Seven years before I started writing this book, after a few years of farming in south-central Pennsylvania – where we could plant tomatoes out in the field in mid- to late May and harvest a good crop until early fall – my wife and I moved to central Maine. We planned on growing tomatoes in the same manner, only with a later planting date to account for the shorter growing season. We were in for a rude tomato awakening.

Our first year in Maine was a perfect storm of tomato woes, which were only a sign of things to come. The first whammy to hit our tomato crop was that spring and summer never really came. Though we planted our tomatoes out in the field according to frost-free dates in early June, it proceeded to rain solidly throughout June and July. By the time our cherry tomatoes were starting to ripen, they were infected with late blight, which is uncommon so far north – the second whammy. We watched our plants wither and die just as they were becoming productive. Barely any tomatoes graced our market stand that year. Since tomatoes are one of the top moneymakers on our farm (as on many small farms), this had a dire effect on our profitability.

But looking at seed catalogs over the winter had restorative effects on our souls, and by the following spring, chalking the failures of the previous year up to fluke weather and diseases, we were ready for another round of growing tomatoes and everything else. So we planted our tomatoes out in the field around the beginning of June, ready for the best. And they *did* do better. Lo and behold, there were tomatoes at our farmers market stand! But . . . a low percentage were top-quality. Our tomato plants grew slowly during Maine's mild days and cool nights, and three of the four horsemen of the humid-region tomato apocalypse – early blight, septoria, and alternaria – developed faster than our plants. The diseases worked their way up from the soil until they reached the ripening fruit, giving them black spots. Big rain events caused splits in the tomatoes. The old-timers tried to assure us that the late blight wouldn't show up again, but we still looked over our shoulders for that fourth horseman to make an appearance and really mess things up. While we had a better year than in 2009, we still ended up with a low rate of salable tomatoes.

We can lament the fact that the produce-buying public is taught to expect airbrushed-looking produce at the grocery store. But it's a fact of life for anyone trying to sell produce that regardless of flavor, the better-looking stuff is going to sell better. And when it comes to tomatoes, splits and spots do tend to make them rot faster. Regardless of how delicious your tomato is, most people are only going to pay top dollar for something that is, if not cosmetically perfect, at least free of major defects. You may be able to get a seconds price for it, but seconds don't pay the bills.

The other discouraging thing we learned about open-field tomato production in central Maine was that the season is very, very short. When you transplant beefsteak tomatoes into the field in early June, you're not picking ripe ones until August. And in an area that usually sees frost in September, that gives you about a month of large-fruited outdoor tomatoes. Of course you can start picking smaller-fruited tomatoes earlier, which makes the season a bit longer. But our customers start asking when we will have big tomatoes in about April, and three to four months is an awfully long wait. I also figure that customers aren't abstaining from tomatoes the whole time they're waiting for ours; they're just buying them from somewhere else. If the customers are already in our market stand, I'd like to be the one selling them tomatoes.

That second mediocre tomato season in our new home led to some thought-provoking conversations during our second winter in Maine. Good thing winter is long up here, giving us plenty of time to chew things over. We had to figure out a way to either improve tomato production or radically change our farm. I distinctly remember having a conversation with my wife, along the lines of "Gee, it seems like all the farms around here making money have hoophouses. Maybe we should build some, too."

It just so happened that at the same time we were puzzling over what to do, I was running the tomato trials at the Johnny's Selected Seeds research farm, half an hour to the south. As I watched the plants grow in my hoophouse tomato trial there, I couldn't help but notice how much slower, later, and rougher everything I was growing in the field at home was.

We decided to build some hoophouses the next season, and we've never looked back. We don't even bother growing large-fruited tomatoes in the field anymore. We still grow some cherry and grape tomatoes in the field, because they are early and fast growing enough to beat the frost and be economically viable. But it's my belief that growing large-fruited tomatoes in the field in central Maine is, in most years, a money-losing proposition. I think most commercial growers in my area, if they looked at how much labor and materials they put into large-fruited field tomato production, and then how much money they earned from selling, would realize they were either breaking even or losing money. And there isn't enough money in the good years to make up for the bad ones. Accordingly, I've witnessed a consistent trend of growers in my area bringing their large-fruited tomato production indoors.

This isn't just a northern issue, though. I've visited older farmers in Florida who in the past were able to reap reliable, profitable tomato crops. But changing weather patterns and rain intensity have led to hit-or-miss profitability. The only way these farmers could regain the level of reliability they needed, and once enjoyed, was to build greenhouses. Of course, greenhouses require a much bigger investment than growing in the field, but the yields and quality are also higher, and the predictability they promise is priceless. As weather patterns continue to change and become more and more unpredictable, protected culture represents a form of crop insurance by allowing growers to control their own climate.

Our first year of growing in hoophouses on my farm was great, but it also showed me how much we didn't know. As part of my job doing trials at Johnny's, I visited a lot of research, trial, and commercial greenhouse growers, all of them better at growing protected culture crops than we were. In learning from these growers, I realized that there is a whole world of protected growing information that is not well known to small- and medium-sized commercial growers. This book is a collection of what I learned putting the techniques from expert professional growers to work on my own farm.

I didn't start out as a greenhouse guy. I learned how to grow by apprenticing on farms in six different states with different growing conditions – Pennsylvania, California, Washington State, Virginia, New York, and Maine – and by starting my own farm. And one thing I've discovered is that the principles of protected growing are the same whether your house covers 100 square feet or 100 acres. The laws of nature apply equally to all sizes of houses, though bigger operations tend to be higher-tech, with subsequently more control over the conditions in their houses. The major difference I found between big operations and small ones is that the bigger growers tend to have a better understanding of the best practices that help their crops succeed.

If you think about how knowledge is shared in farming, this is not surprising. Know-how tends to be passed from generation to generation, or people learn the craft directly from someone else, like I did as an apprentice. I do believe that experience is the best way to learn such an all-encompassing skill set as growing.

The disadvantage to this method, however, is that you don't get a survey of available techniques and technology, as you would in, say, a college course. Even if you wanted to go to school to learn greenhouse growing, the options in North America are very limited, with only a few schools teaching such techniques. Combine this with the fact that much of the cutting-edge research and development for protected culture is done in the Netherlands (and that larger growers don't tend to publicize their best information due to competition), and it's not surprising that most smaller growers aren't aware of the best practices.

This book is written to help vegetable growers get the most out of their investment in protected growing space.

Whether you're currently growing in a greenhouse or hoophouse or not, this book can help acquaint you with the best practices to maximize the potential of the most popular protected crops.

For those who do not have hoophouses or greenhouses yet, or who want to expand their protected growing space, this book can help you decide what crops to grow, what features make a suitable structure for the production of certain crops, and how to manage those crops from seed to post-harvest. Though I had commercial growers like myself in mind while writing, any gardener serious enough to put up a hoophouse or greenhouse will want the same advantages.

Why These Eight Crops

This book focuses on eight crops – tomatoes, peppers, cucumbers, eggplant, lettuce, greens, microgreens, and herbs – because these are the most reliable crops to grow and offer the highest return on investment in protected growing space. They enjoy a high year-round demand and need to be sold fresh, plus their yield and price per pound are high enough to justify taking up precious protected growing real estate. These eight crops are so important to protected growing that they have evolved their own specialized production methods and varieties separate from field growing.

All of the crops in this book fit into one of two cropping models. The vining/fruiting crops – tomatoes, cucumbers, peppers, and eggplant – are called long crops because they use one or a few long cropping periods over the course of a season. A greenhouse pepper crop, for example, can be seeded as early as October, transplanted into the greenhouse in early December, and harvested from mid-March until the following November.

With long crops, you can plant seeds for the next crop before the previous crop is even out of the ground. The plants may be over a year old by the time they're removed from the greenhouse. The only time the greenhouse isn't growing is a cleanout window, typically a month or so around the New Year. The long harvest period makes up for a long establishment period before anything is harvested. Pepper plants, for example, take a relatively long time to get to harvest. They may grow in the greenhouse without being harvested from December until March. The fact that they can be harvested for close to nine months pays for the three months or so that they grow before harvest.

The leafy crops – lettuce, greens, microgreens, and herbs – are called short crops because the time between propagation and harvest is so short. Lettuce is a good example of the short greenhouse crop model. In a greenhouse, lettuce can be seeded continually throughout the year, with the crop taking approximately eight weeks from seed to harvest. Young plants may spend up to half that time densely spaced in propagation, and only tie up production greenhouse space for four weeks before being harvested and making room for another round of plants.

No single lettuce crop will be worth as much as a tomato crop, but the greenhouse may accommodate ten or eleven lettuce crops over the course of a year. Microgreens have even shorter cycles, being ready in as little as seven to ten days. The name of the game with any short crop is to minimize the amount of time from propagation to harvest, in order to pack as many harvests into a season as possible.

One way to drive the point home about why I chose these eight crops instead of others is to consider the potato. I know a market grower who once grew potatoes in his hoophouse. He told me, "They grew great! Never grow potatoes in a hoophouse." What he meant was, potatoes will grow well in protected culture, but they're not profitable.

Potatoes are a poor choice for a greenhouse crop in part because they store well, so they are available locally for much of the year without having to be constantly produced fresh. They also bring a relatively low price per pound. The third nail in the coffin is that potatoes tie up a lot of time in the greenhouse for only a single harvest. This is in contrast with their solanaceous cousins: tomatoes, peppers, and eggplant. Let's say that you have potatoes and peppers growing in your greenhouse, and they both take the same number of days to mature. When the day comes that both crops are ready to harvest, you dig your potatoes once and that's the only harvest. You have to replant your greenhouse to something else.

The day you start picking peppers, however, you can continue picking them for the next eight months or so if you want to.

I'm not saying that potatoes can't be a great crop for a home grower who has extra space in a hoophouse and wants some extra-early potatoes. But it would be very difficult for a commercial grower to justify as a protected crop. There are just a lot of other more profitable things a commercial grower could devote the space to.

Let's say you want to grow a crop that's not on our list in a greenhouse or hoophouse. Maybe you have a really great year-round market for radishes. Though radishes are not among the most common greenhouse crops, they can be profitable. Since radishes are so fast growing, they could fit into the short crop production model like lettuce and greens. Many radishes are a month or so from seeding to harvest, so they can fit into roughly the same harvest schedule as lettuce, with ten to twelve crops a year. You could grow a couple of short crops of radishes when the greenhouse is out of production from another crop, or mix them in with other cool-weather crops. Or you could experiment with growing them in flats on unused bench tops. There are a lot of possibilities.

It's worth evaluating the many other vegetable crops that are not covered in this book to decide for yourself if they are a good fit for your growing style or not. The crops covered in-depth in this book are the most obvious ones to consider if you want to run a farm or greenhouse business, or help feed yourself with protected culture. That doesn't mean there aren't other crops that can be profitable in a greenhouse – just that most of those crops will fit one of the above profitable crop models.

How to Use This Book

This book is a collection of best practices for the main protected culture crops that are different from raising crops in the field. Although I've included an appendix on pests and diseases, and some information about siting your greenhouse or hoophouse, this book is not primarily about those subjects. What it does do is give you the basics that make the Dutch system of protected growing so much more productive than field growing, and show you how these techniques can be adapted to unheated and sustainable growing.

I've intended this book to be just as accessible to beginning growers as to experienced ones, whether you're trying to decide if you should build a hoophouse for the first time or have been growing with a greenhouse for twenty years.

The book has a lot of rules, but they are not hard and fast. Horticulture is a game of making plans based on what you expect to happen, but success is determined by how well you react when things don't go according to plan.

No one will ever be able to tell you exactly how a season will go, so no one will be able to tell you exactly what to do beforehand. Individual best practices, picked and chosen based on your own priorities as a grower, will eventually turn into a management strategy. Familiarity with the best practices will give you an idea of how to deal with challenges before they manifest themselves. Having a strategy is the difference between being proactive and reactive. It allows you to see the big picture; rather than only thinking about what has to happen today or tomorrow, or just focusing on light, water, or temperature, it will help you consider all the factors of protected growing. Then when the weather does something unexpected or pests show up, you're prepared to deal with the situation.

I hope this book will give you, as a grower, the tools to develop a management strategy all your own. One thing I find myself saying a lot is that there are as many ways to grow as there are growers. Within the best practices presented in this book, there's a lot of room to develop a management style that suits your personality and priorities as a grower and the circumstances you're growing in.

The ideas in this book about how to manage protected culture vegetable crops should be seen as an à la carte menu of management techniques, where you can take those that appeal to your growing style and leave those that don't. Growers should come away with a better idea of why their plants are behaving in a certain way, and how to influence them to do what they want. Take the techniques that appeal to your level of technology and expertise, and form your own management style that reflects your philosophy as a grower.

After trying the techniques I learned in larger greenhouses on my own farm, I can say with confidence that they all worked at least to some degree on my own small scale, which is why this book exists. The same cannot generally be said for practices on the largest farms in field agriculture, where large acreages, economies of scale, machinery, and chemicals are central to conventional production.

What I found is that the basis of high production in large Dutch-style greenhouses comes down to plant management, climate management, and efficiency. Granted, a small grower is not going to invest in a motorized harvest cart with scissors lift, but the higher price that a small operation gets for its produce should make up for lower efficiency.

Recently someone asked me, "There's so much emphasis on having the earliest tomatoes; what are your suggestions for having the *last* tomatoes at market?" My suggestion was simple: Build a greenhouse or hoophouse. Regardless of where they are grown, the last straggling tomatoes from the field won't be as good as those from the height of the season, since they mature under marginal conditions. Even the simplest hoophouse will extend good tomato weather another few weeks.

I look at early and late tomatoes as an analogy for our local food system in general. One of the changes that I would like to see in the world is for more of our food to be produced closer to where it is eaten. One of the many ways to make progress on this goal is to extend local food seasons, so more crops can be produced over more of the year everywhere.

There will always be challenges in growing, because pests, diseases, weather, varieties, and market demands change from year to year, but that's one of the things that keeps growing interesting for me. I hope protected growing can help by taking out some of the unpredictability and giving growers a greater degree of control over the weather and their plants – at least in their hoophouses and greenhouses.

It's now been years since our rude tomato awakening. If we had stayed in Pennsylvania or some other place with better tomato weather, I probably never would have been pulled to protected growing, or at least not as strongly. We try new things every year, and we've had good luck with many of the things we've tried. This book is a collection of the things that worked that now make up the backbone of our system.

PART ONE

Protected Culture Basics

CHAPTER ONE

Why Protected Culture?

The term *protected culture* refers to the practice of growing plants in a structure designed to protect them from environmental stresses and improve the growing environment. Though it's not heard often in North America, *protected culture* is commonly used in Europe to describe the full range of protected growing, from the simplest unheated hoophouse to the fanciest heated greenhouse. I hope to help the term gain more usage in the Americas as protected culture takes on an increasingly important role in the food production system. It's a handy term to have when you want to talk about protected growing in all its forms.

The Dutch Influence

You may not be surprised to learn that the United States led the world in agricultural exports in 2015 (the year for which the most complete statistics were available at the time of this writing). What might surprise you is that the Netherlands ranked second. Area-wise, this is like West Virginia coming in second to the entire rest of the United States, except that West Virginia is approximately a third *larger* than the Netherlands. So it's really like two-thirds of West Virginia coming in second to the rest of the US.

To put it another way, the 134th largest country in the world, with the 65th largest population, came in second to the 3rd largest country by both area and population. How? Advanced greenhouse techniques are a big part of the reason why the Dutch have such an oversized agricultural output, especially in the categories of vegetables and ornamentals, which they lead.

When I learned about the Dutch leadership in protected culture, I wanted to know how many of the Dutch greenhouse innovations would work on smaller farms. I decided to find out by putting them to work on my own farm and in the trials at Johnny's to see how well they fared in lower-tech situations. The short answer is: Most of them work really well in smaller settings.

Figure 1.1. Dutch eggplant in my local grocery store in Maine.

Figure 1.2. Some of the author's hoophouses in winter.

One of the foundations of Dutch greenhouse growing is planting crops as densely as possible. On my farm, we were able to go from growing roughly 1 ton of tomatoes in a 30 × 48-foot (9 × 15 m) hoophouse, to 2 tons, mainly by switching from single rows to double, doubling the number of plants we put in the hoophouse and therefore also the yield. But it was other techniques, like using both fans and pruning to promote good airflow, and precise temperature and humidity management to keep the crop healthy, that allowed us to plant that densely without inviting problems.

These techniques can help make smaller farms more productive, profitable, and competitive with produce on the commodity market. Though they have been used to maximize productivity in some of the largest greenhouses in the world, they are scalable and can increase productivity on smaller farms without sacrificing quality and sustainability. In fact, by increasing the yield per unit of input, they may be able to increase the sustainability of many operations.

The basics of the Dutch greenhouse system are:

Spacing: Maximizing the number of plants that can be grown in a given area.

Climate: Managing temperature and humidity not only for fast growth, but also for the type of growth you want (say, leafy versus fruiting).

Plant Husbandry: Using specific greenhouse growing practices to care for plants in a manner that promotes their productivity.

Variety: Choosing varieties that are adapted to protected conditions.

Structures: Using the right kind of structure for your area and what you are growing.
Efficiency: Operating with as little waste as possible.

Hoophouses and greenhouses are expensive to build; the area inside is precious real estate. By learning from these techniques, you can get the most out of your protected space.

The Importance of Protected Growing to Local Food Systems

Protected horticulture is one of the most important ways for local food producers to compete with industrial agriculture. Consumers are used to a food system that can deliver them almost any food at any time of the year. Industrial agriculture identifies the best place to grow a particular crop, grows it there, and ships it everywhere else, taking advantage of ideal production conditions, economies of scale, and cheap transportation. Our food system has swung a long way in the direction of centralization, and protected culture is an important part of re-regionalization.

Local food producers face two major challenges when it comes to taking market share back from industrial agriculture. The first is geographic: In much of the country, most produce is seasonal and not grown year-round, so consumers – especially in colder climates – have few local options in the winter for certain popular vegetables. The second challenge relates to production and marketing in that many people see food as a commodity to be purchased as cheaply as possible. So local food producers have to either figure out how to match the price set by the industrial food system, or differentiate their food so consumers will pay more for it.

My opinion is that we reached the high-water mark for industrial food in the United States sometime in the 1970s or '80s. Consider that if you go back far enough, almost all food was local; long-range transport systems did not exist for perishables. The imported food that was available, like sugar and spices, was expensive and nonperishable. Fast-forward from 1780 to 1980. Refrigeration technology and the interstate transport system made it possible to deliver produce anywhere in the country (and beyond).

By the late twentieth century, most food production was centralized in areas with the most favorable outdoor conditions for a given crop. Take lettuce, for example. Today somewhere between 50 and 80 percent of the nation's lettuce comes from a single valley in California, the Salinas. Why? Because it's one of the best places in the world to grow lettuce. If you are a lettuce plant, you will be very happy growing in the Salinas Valley most of the year (in the wintertime, lettuce production switches to Arizona). You might think, so what if most of the country's lettuce is grown in one place? It makes great sense, except when you consider both the value of widely distributed agricultural production, and the cost of industrialized food production.

The industrial production model, for example, incurs a number of costs that are socialized (the environmental costs of monocropping and long-range transportation; the social costs of relying on undocumented laborers), whereas the profits are privatized. If all costs were taken into consideration, locally produced lettuce might well be cheaper. However, in our system that only recognizes monetary costs, industrial production is "cheaper." So we find ourselves in a situation where most states are produce importers, while only a few states are huge produce exporters, led by California and Florida.

In the past few decades, however, even with minimal awareness of the environmental impacts of industrialized farming, consumers have noticed changes in their food. Everyone knows that the term *grocery store tomato* refers to a cosmetically perfect, hard, flavorless tomato. The tomato may be the poster child for bland produce, but it has become evident in many crops that efforts in breeding and growing were for the purposes of easier shipping, not better flavor. National produce recalls for *Listeria* and other pathogens have only highlighted the fact that massive amounts of produce come from single sources.

Whatever the reason, whether backlash against industrialized food, affinity for the advantages of local food, or both, the local food movement has begun to gain momentum. The most obvious indication of this momentum is the number of farmers markets. The USDA started keeping track in 1994, when they counted 1,755. Twenty years later, in 2014, they counted 8,268. That's almost a fivefold increase over two decades. So clearly consumers are interested in buying locally produced foods. But as mentioned above, most local growers can only produce for a limited part of the year, while the industrial food system can offer anything anytime. Many farmers market shoppers come out to get what is seasonally available, and then head over to the grocery store for everything else.

Increasing protected culture growing is the most important way to maintain the momentum of the local produce movement. Though there's now a market for local food, local growers draw the market when their food is in season and lose it when it goes out. One of the best ways to grow local food sales is to grow the length of the season through protected culture.

Out-of-season industrial production is vulnerable to competition from local protected culture growers. Florida's winter field-grown tomato market is collapsing. If you want proof of this, check out the winter tomato selection at your local grocery store, which used to be dominated by green-harvested gas-ripened field tomatoes. Now over 70 percent of winter tomatoes for fresh eating are grown in greenhouses. And understandably so: Since most greenhouse tomatoes are harvested at least partially ripe, they deserve the reputation of being at least better quality than winter field tomatoes.

The trend toward greenhouse tomatoes is not surprising when, as one breeder told me, the goal of winter field tomato breeding is to produce a tomato that "could smash a windshield." If you have any doubt about the production shortcuts that lead to the poor flavor and quality in winter field tomatoes, read *Tomatoland* by Barry Estabrook. I thought I knew about the produce industry, but this book shocked me.

Protected culture is the only way for many local growers to compete with out-of-season production from industrial agriculture. Since not everyone is willing to pay a premium for local food, even if it does taste better and is fresher, the challenge is how to produce local food out of season at a price that is competitive. As a local food proponent, I know that people will always want to buy produce out of season. I'd rather see the dollars for out-of-season production go to a local grower.

Now on to combating the perception that local food is more expensive. At my farmers market, our prices are frequently the same as or similar to those at local grocery stores, and we've used signage in our market stand to point this out to customers. But the larger dynamic that needs to be overcome is that billions of dollars of farm subsidies go to commodity crops – corn and soy in particular – that are processed into junk food. This puts any type of vegetables, local, organic, or not, out of reach for many household budgets, and makes the worst types of junk food the most affordable option for the poorest people, who need to maximize the number of calories they get out of each dollar. Ending agricultural subsidies is a subject for another book, but I do believe that the impact of making junk food artificially cheap is that vegetables of all types seem more expensive.

I see an opportunity for local growers to take advantage of the fact that industrial food has such high transportation costs associated with it. The lack of such costs needs to make up for the higher cost of production associated with most smaller operations. Plus, our carbon footprint would be smaller if local growers pocketed more of the price of produce, instead of it paying for interstate transport.

Produce wholesalers don't like the current long industrial supply chain that much, either. It's cheap and efficient, but it doesn't produce a high-quality product. Many would prefer to offer tastier, fresher local produce, but wholesalers' need for a reliable supply has long been a barrier for seasonal, local growers. Greenhouse production is a way for local producers to finally remove that barrier, since production becomes more dependable without the vagaries of weather.

In an era of foodborne illness scares, wholesalers also appreciate the fact that food from protected agriculture is produced in a controlled environment, which should be clean and free of contamination. Ultimately, local protected culture growers can offer food that is higher-quality than anything industrial food has to offer.

Future Trends

There are some threats on the horizon for the industrial food system, which will increasingly favor local production. Weather disruptions associated with climate change are one. In North America, for example, a prolonged drought in California is currently jeopardizing the most important vegetable production region in the country. Not having enough water to maintain prior production levels, as well as scarcity forcing prices up, both represent an opportunity for local growers.

The gradual increase in the price of fossil fuels could also diminish industrial production. Cheap fossil fuels are the foundation of the industrial food system and its ability to ship food anywhere in the world. Though they may continue to be relatively cheap now, the inevitable increase in transportation costs will increase the cost of industrial food. In the long run, this will favor local growers with shorter supply chains. Improvements in energy efficiency and green technology mean that local greenhouse growers can grow out-of-season produce efficiently, closer to the consumers.

On the farms and greenhouses that I have visited, growers are investing in protected culture production at a faster rate than they are investing in new field production. This trend, coupled with the fact that the Natural Resources Conservation Service has funded over thirteen thousand hoophouses to date in all fifty states, means that there is a rapidly growing amount of protected culture growing space on farms everywhere.

Figure 1.3. The author working on the end wall vents of one of his hoophouses.

The rise in demand for local food has created an opportunity that farmers can meet through protected agriculture, which has an especially important role to play in the local food movement. Even though year-round field growing is impossible in most of the country, people still need to eat 365 days a year. So the choices are either to let local produce dwindle down to storage crops over the winter in most areas, or to extend the season. I know there are some dedicated locavores who prefer to subsist as much as possible on food that can be stored over the winter, much as our ancestors did. But with people used to the wide variety of produce that's available year-round from the grocery store, not everyone is going to go without fresh produce in the cold months.

Now that farmers markets have reestablished themselves in many communities, extending the seasonality of the foods available from local production is the next step for increasing the consumption of locally produced foods. We have made great progress in increasing the accessibility of local foods through farmers markets, CSAs, food hubs, and local wholesaling. And with many foods, the season can be extended through refrigeration, freezing, or storage. With the eight crops I focus on in this book, however, if no one is extending the season at the end of the local field-growing season, the dollars spent on local food will go back to industrial agriculture. This is why raising the quality and quantity of season-extension crops is the next step in the revitalization of our local food system.

Not only do I support local food systems because I am a local grower; I am a local grower *because* I support the idea of a robust local food system. I think it would be great if more of the food we consumed were produced locally. For one thing, local food systems are resilient. In a centralized food system, if bad weather impacts an important production area it can cause widespread disruption. The disruption doesn't even have to be in the production area to have an effect. Lettuce prices are seldom higher in the East than when the Mississippi floods, preventing bridge travel and transport from the Salinas Valley. This highlights both the extreme centralization of production and the importance of transport to the industrial food system. When lettuce production is distributed over the country, however – whether in field or protected culture – localized weather and climate disruptions don't have as big an impact.

Local food also means fewer food miles. Some people may disagree and say it doesn't matter, that fossil fuel usage will come in the form of greenhouse heating if not from the truck transporting the food. But greenhouse heating systems are getting more efficient all the time, to the point where it's now possible to have an off-grid, carbon-neutral or fossil-fuel-free greenhouse. Over time, the increasing cost of fossil fuels will favor efficient greenhouse production over long-range transportation of produce. Every part of the industrial food chain, from plant breeding to harvesting, is made to support shipping produce long distances. If produce didn't have to go as far, not as many corners would need to be cut, resulting in better-tasting, more nutritious food.

Greenhouses and Sustainability

As greenhouses become more common, I hear more people questioning the amount of resources that go into greenhouse growing: "Is greenhouse growing sustainable?" The answer is . . . it depends.

The level of sustainability varies from one greenhouse to another, so there is no cut-and-dried, yes-or-no answer to whether greenhouse production in general is sustainable. The answer depends both on the object of comparison and on your definition of sustainability. For instance, if you compare the amount of resources that go into producing a head of field lettuce in California in

Figure 1.4. This experimental off-grid solar-powered greenhouse is at the University of Arizona in Tucson. All the electrical components are as efficient as possible to make the most of the electricity that is generated.

June with the resources that go into greenhouse lettuce in Maine in January, field production comes out on top for energy use. However, greenhouse production is more efficient in its water usage than field production in an arid environment like California. A better comparison would be between greenhouse winter lettuce production located close to the consumer and winter lettuce grown in the field in Arizona (the main winter lettuce production area in the United States), factoring in the miles of transport necessary to reach the consumer.

I'm not even going to get into how much tractor fuel and water was used to grow the lettuce in Arizona, plus all the miles on a semi to get it across the country, versus how much greenhouse energy it takes to grow and deliver a head of lettuce locally in Maine. For one thing, these numbers change all the time depending on how far the field lettuce has to travel, what kind of heating system was used for the greenhouse, and so on. It's also difficult to calculate how much more beneficial it is that food dollars spent locally circulate longer in the local community than those spent with national corporations. The bottom line is that greenhouse growing is getting more sustainable all the time, and, if it isn't already, it has the potential to be more efficient than centralized industrial field production. The technology already exists for an off-grid, solar-powered, geothermal greenhouse with net-zero carbon emissions. And the technology is improving all the time. The economic pressure for growers to be as efficient as possible is great, since for many greenhouses, heating is the second biggest expense after labor.

Water and Fertilizer Use

Greenhouses tend to be efficient users of water and fertilizer. Thrifty greenhouse production is based on giving the plants the right amount – no more than they need – of fertilizer and water. When compared with open-field irrigation methods in arid parts of the country, water usage is likely more efficient in greenhouse production. Relatively little water is lost to evaporation in greenhouses, so a high percentage of irrigation water ends up doing its job growing the plant. When you take into account that much of the United States' centralized field production takes place in parts of the country where water is least abundant, greenhouse water efficiency looks even better.

Energy Use

Heating a greenhouse to grow warm-weather crops in the middle of winter is resource-intensive, which is why most long-season greenhouse growers both start plants and end the crop during a cold part of the year, but avoid the coldest, darkest part. It is in the interest of growers to minimize the amount of energy they use. In fact, a lot of greenhouses went out of business during the energy crisis of the 1970s, when fuel prices spiked over a short time. That winnowed out the least efficient operations, and efficiency has been a priority ever since. Improvements in efficiency and new heating sources mean that greenhouses are now more efficient than ever.

Hyper-efficient technology is the future of greenhouse growing, and the gradual rise in fossil fuel prices due to dwindling supplies will only speed up the arrival

Figure 1.5. Sunset by the hoophouses.

of this future. So being efficient will become even more lucrative. Growing interest from the public in how their food is produced will also spur more energy efficiency in greenhouse growing. Smart growers – and growers who sell into their local market especially – will take advantage by turning this interest into a marketing opportunity. The sustainability of their produce can become a selling point, along with its flavor and other attributes that set it apart from industrial food, helping brand their produce as sustainable to those in the local community.

Grower Benefits of Protected Agriculture

In addition to benefits for the consumer, the environment, and the local economy, protected agriculture offers many benefits to growers beyond simply allowing them to be the first farmer to market with any particular crop – though this is important as well.

Earliness

I once took a business trip to visit greenhouses in Canada. One of the crops I looked at was eggplant grown in big commercial greenhouses. When I got back home I went to the grocery store in central Maine close to where I live (I don't actually buy a lot of produce at the grocery store, because I try to grow as much of it myself as possible, but when I'm there I always ogle the produce to keep up with trends and prices). There in a display I saw some beautiful greenhouse eggplant that looked just like the varieties I had been looking at the day before in Canada and thought it must be from just over the border. Upon closer inspection it turned out to be from Holland! I can't say I was so surprised that I dropped the eggplant, but it did make me wonder why eggplant was traveling to Maine on a plane all the way from Holland when I knew the exact same stuff was growing a couple of hours away in Canada, if not closer. What it really made me think was, *I should be the one growing the eggplant for my local grocery store.*

If you ask growers why they grow in a hoophouse or greenhouse, having crops both earlier and later than is possible with field production is probably one of the first benefits they will mention. It's almost too obvious to note. Protected culture turns pepper season from three months into however-long-you-want-it-to-be. So long as peppers are profitable for you (do the math and make sure!), it gives you that much more time to sell a profitable crop. And since peppers are available at the grocery store from somewhere in the world every single day of the year, as far as I'm concerned it might as well be you getting paid for growing them.

Everyone wants to be first, even people who aren't growing for profit. The home gardener has the satisfaction of being the first in the neighborhood to pick a ripe tomato. President Thomas Jefferson used to hold a yearly competition with his friends to see who could pick the earliest peas. But being first has a particular significance for commercial growers. Growers who sell direct at farmers markets understand how having the first of anything will get people in your stand in the spring, and if you can get them started buying from you they will likely be your customers for the rest of the season. Whether you're filling a market stand or a CSA box, earliness, consistency, and diversity of selection keeps customers. One of the main complaints I hear about CSAs from customers is getting the same thing week after week. But the earlier and later you can produce any given crop, the more diverse your offerings will be, since you'll be able to offer a wider selection over a longer period of time.

Being first and last can also help you land and keep market and wholesale accounts. Maybe your goal is to wholesale cucumbers to your local health food store. If you're the first one to offer them cukes in the spring, they're likely to become your customer. My local grocery store chain probably buys eggplant from Holland because they can start early and provide a consistent supply throughout a long season. If I could offer them the same thing, they'd probably buy from me.

Protection: Consistency, Predictability, Crop Insurance

Hand in hand with earliness comes predictability, because it doesn't do you much good to get a market early in the season only to lose it later on because you can't deliver consistently. This is where the *protection* in protected agriculture comes in. We all know Mother Nature can be fickle and temperamental. Protection comes in handy when it comes to weathering the hailstorms, torrential rains, droughts, and other adverse weather that is bound to come your way sooner or later. Depending on what level of control your structure affords you, you may be able to even out bigger bumps in the road, such as freezing or triple-digit (in the 40s, if you use metric) temperatures.

I remember talking to one grower who grew both field and greenhouse crops on his farm. After weathering a tropical storm, he said his outdoor crops in river-bottom fields were ruined. But he still had a house full of greenhouse lettuce to put on his farm stand, which he'd harvested while watching his pumpkins float away down the river.

Weather patterns are changing, becoming more and more unpredictable. There are also an increasing number of extreme weather events. I know a lot of growers who have put up structures as a form of insurance to shield their crops from capricious weather.

Higher Yield and Packout

Every grower knows: It doesn't matter whether you can grow it or not if you can't sell it. And if produce is damaged in almost any way – from bruises to splits – you may not be able to sell it, and you almost certainly can't get full price for it. Protected culture can help to both increase yields and increase the amount of marketable produce.

Figure 1.6. Greens in a greenhouse.

In later chapters, I will show you how to manipulate the environment to optimize plant growth. In addition to increased yield, this should also lead to an increased rate of top-quality produce, as the plants can be protected from bad weather, some pests, and some diseases as part of controlling the environment.

Efficiency

The term *force multiplier* refers to something that amplifies the impact of an effort. Protected culture is a force multiplier for agriculture. Though it requires a bigger up-front investment than does plowing a field and growing vegetables in the open, when used properly, protected culture increases the value of the labor and resources put into it. For example, high-tech tomato greenhouses can have yields in the range of twenty times greater than a field of the same area.

This is the high-tech growing model: high investment, high return. Protected agriculture makes the most of the heat, light, fertility, labor, and plants that you invest. Whether you want to invest a lot or a little, protected growing can increase the returns on what you put in.

Pest and Disease Control

I wish I could say that protected growing would make your pest and disease problems go away. But the fairest thing to say about pest and disease control in protected culture is that you tend to trade certain problems of the field for other problems more common to growing indoors. Still, an advantage of protected growing is that one of the tools for managing pests and diseases is the structure itself; the more control you have over the environment, the better you can control some pests and diseases.

From the humblest high tunnel to the fanciest glass house, one advantage of protected growing applies equally: the ability to keep the leaves dry. Almost all crops suffer from afflictions that are worse with wet foliage. So simply by going under cover, growers avoid all sorts of rain-related problems. For a brief rundown of common pests and diseases, as well as strategies for dealing with them, please see appendix B.

Improved Flavor

A lot of what affects the flavor of a vegetable is up to the grower. People tend to refer to certain varieties as good tasting or bad tasting. But the reality is that flavor is the result of the way a vegetable is grown as much as of its innate potential. I've seen some growers kill varieties with good flavor potential, and others get varieties with average potential to taste really good with careful growing for flavor. And though it's not always true, a lot of practices that improve yield hurt flavor, and vice versa. So just as protected culture isn't a magic bullet for pests and diseases, it also doesn't automatically translate into better flavor. But for growers who want to prioritize flavor, it gives them some extra tools in the toolbox for growing tasty vegetables.

For one thing, protected culture can take advantage of some varieties that are hard or impossible to grow in field production, like seedless, thin-skinned cucumbers and cluster tomatoes. Additionally, excluding the rain prevents the flavor from getting washed out of tomatoes. Whereas field growers' options are limited to picking aggressively before a rainstorm or crossing their fingers that their tomatoes won't split, protected growers can pick their tomatoes whenever they're ready. Plus there are other tricks unique to protected growing, like growing the crop with a high electrical conductivity (EC) – in salty soil or with moderate water stress – to improve flavor. Developing a following is all about flavor for local growers. When you sell direct, standing behind your produce, people will come back to your market stand or CSA if you have the best-tasting produce.

In the next chapter, we'll look at the different types of protected culture structures – from those without heat (hoophouses) to those with (greenhouses) – since the design and features of each will determine the amount of control growers have over their crops, and subsequently how many benefits growers will receive from bringing their growing indoors.

CHAPTER TWO

Protected Culture Structures and Their Features

The term *greenhouse* is usually applied to heated structures, whereas *hoophouse* or *high tunnel* is used for unheated structures. In this book, I'll use the term *protected culture* to refer generally to all forms of protected growing.

The principles of protected growing don't change much, but the technology is changing rapidly, so wherever possible I have left out brand names when talking about products and equipment. Growers will be better served by getting in touch with the vendors of any particular service or technology and finding out what the best solution currently is for their situation. I'm not going to tell you what to buy, but I am going to help you determine what questions to ask so you can figure out what to buy.

When pricing out options for greenhouse products and services, heed the following best advice (which can also be applied elsewhere in life): You can get it fast and cheap but not good, good and fast but not cheap, or cheap and good but not fast. Translation: Start sourcing and getting quotes as far ahead of time as possible. If you need anything at the last minute, you are going to pay for it.

Get as many quotes as possible, let the salespeople know you're getting quotes from other companies, and then look at the features of the various quotes, which might explain differences in pricing. Though it's usually true that you get what you pay for, you may be able to save yourself some money by comparing features to make sure you're comparing apples to apples, and then eliminating overpriced options based on price and features.

Choosing a Structure Based on Type of Crop

When you're thinking about building a structure for protected culture, one of the first questions to ask is: What type of crops will you be producing there? Some greenhouse experts and consultants advocate a one-greenhouse/one-species approach. They think that to really optimize the conditions in a protected culture structure, you must devote the entire space to a single species so that you can cater to that crop's needs as much as possible. And there is an element of truth in this. If you're growing more than one species in the same greenhouse, you have to choose to either optimize the conditions for one of the crops or compromise among the requirements of all crops concerned. What this approach does not take into account is the value of crop diversity to many smaller growers. Depending on their markets, many growers can make more money and serve their customers better, or simply enhance their lives, by having a more diverse growing environment.

Still, crop groups should be taken into account when building a new structure. Broadly speaking, the leafy-crop group has lower requirements for fertility, light, and temperature than does the vining/fruiting group, since it takes a lot more inputs for a tomato plant, for example, to crank out a bunch of fruit in addition to leaves.

Vining crops also grow much taller than low-growing greens. Though in general greenhouses are trending taller, modern leafy-crop greenhouses tend to be in the range of 9 to 10 feet (2.7–3 m), while modern vining-crop houses are in the range of 18 feet (6 m) to the gutter. Originally, greenhouses were constructed to be as short as possible in order to minimize the amount of air that had to be heated. But we've since realized that it takes less energy to maintain the temperature of a larger air mass, which is more buffered to temperature changes. However, if you're planning on growing a long-season vining crop that isn't lowered and leaned (see chapter 6), such as a pepper crop, you may end up actually using all 18 feet up to the gutter. Height is another reason why it can be difficult to mix greens and fruiting crops: The vining crops are so much taller that they tend to shade the lower-growing greens.

Keep these basic considerations in mind as you read further into this chapter about the types of structures, materials, and other features that may or may not be applicable to the crops you want to grow.

Siting East–West Versus North–South

One of the most basic considerations underlying success in protected culture is the location of the structure. When considering whether to run a greenhouse or hoophouse north–south or east–west, there are a number of dynamics to consider. As you get farther away from the equator, greenhouses oriented east–west allow in more light during the winter. This is because when the sun is low, it will strike the side of an east–west structure more directly and more light will penetrate.

On the other hand, depending on the crop you're growing, an east–west orientation can create a lot of shading inside the structure. If you're growing tall vining crops and the rows run the length of the greenhouse, the southernmost row will receive a disproportionate amount of light and throw shade on the rest of the house. Lighting and shading may be more even for crops that run north–south, because each row gets the sun overhead for a portion of the day. So, to contradict the earlier advice about light penetration, I have always built my structures north–south despite the fact that I'm pretty far north.

The limitations of your property should also be taken into account. Don't site your structure awkwardly just to make a certain orientation work. If your property is a square oriented northeast–southwest and you need to use all of it, just go with a northeast–southwest orientation even though it doesn't use the cardinal directions.

The Scale of Protected Culture Technology from Low to High

In field growing, growers try to fit the crop to the environment. Varieties are chosen based on anticipated weather and disease conditions. The opposite is true in protected growing, where it's the job of the grower to fit the growing conditions to the crop. The ability to control the weather in your own protected environment is priceless when it comes to maximizing greenhouse vegetable crops. Knowing how to use that increased level of control is what this book is all about.

Hoophouses and greenhouses are frequently referred to as low-tech or high-tech. These distinctions refer to a scale from a lower to a higher degree of technology and control over the growing environment. They are useful distinctions to make, because management strategies differ based on how much control you have over the environment.

The bottom line is that higher tech = higher cost = higher degree of control. Growers who want it will pay more for more control, because it makes things possible that aren't with a lower level of environmental control, like harvesting ripe bell peppers in Canada in November. To explore the scale of environmental control offered by each option, we'll look at the possibilities, starting with the lowest of low-tech going up to the highest of high-tech. What you choose to build will depend on your budget, the size of your farm, what crop or crops you want to grow, and the geographic location of your farm.

Field Tunnels and Caterpillar Tunnels

Though greenhouses have gotten very complex, even the most basic unheated structures provide many advantages over growing in the field. Just having a piece of clear plastic over a crop fulfills the most basic function of protected growing, which is to increase the length of the season by trapping solar energy. Another huge benefit of covering the crop is that it keeps precipitation off. This is an obvious advantage when it comes to hail, but when you consider how many foliar and even root diseases are caused by moisture and humidity, keeping rain off is also a plus.

The simplest high tunnel is just a piece of plastic pulled over the crop without any side or end walls. Whether built from a kit or scratch, the basic tunnel involves posts that are driven into the ground, with hoops or bows arched overhead. Corrosion-resistant steel pipe is most commonly used, though other materials are also options – including the weaker and cheaper PVC pipe. Usually a pipe called a purlin runs end-to-end overhead, tying all the bows together and giving the structure stability. In some temporary structures, rope may be used for a purlin instead of a pipe.

One common design uses ropes attached to the base of each bow, which are then laced over the plastic to anchor it to the structure. When built as stand-alone tunnels, these are frequently called caterpillar tunnels, as the ropes crisscrossing over the top of the tunnels give them a segmented look similar to a caterpillar. Caterpillar tunnels can be advantageous for the grower since they allow for variable side ventilation without the expense of installing rollup sides. The plastic can be pushed up by hand under the ropes to ventilate the sides, a feature that is more time consuming but cheaper to build.

When a larger growing area needs to be protected, tunnels are joined together at the ground posts. Tunnels joined together in this manner are called gutter-connected, because the place where the two tunnels join together forms a trough (which may or may not actually function as a gutter depending on the design). Tunnels can also be referred to as bays, especially when they are gutter-connected, as in a multibay structure.

The reason that growers gutter-connect their tunnels, instead of lining a bunch of individual tunnels up next to each other, is because it's much more space-efficient to connect them. Tunnels that are built side by side without being connected need to be spaced out so they don't cast shadows on each other – usually a minimum of 10 feet (3 m) between tunnels. If they're gutter-connected, however, the whole area can be covered with no open space between the tunnels. It's also much more efficient to ventilate a single large structure than multiple smaller ones.

The advantage of field and caterpillar tunnels is that they're relatively cheap, are quick and easy to build, and can still give many of the benefits of a fully enclosed hoophouse when it comes to extending the season by a few weeks on either end. Worldwide, low-tech tunnels cover thousands of acres because they are the simplest, cheapest, and quickest type of structures to install. This includes tens of thousands of acres in warmer areas, where a simple covering may be all that's needed to grow through the winter.

The disadvantage of this type of structure is that it offers no frost protection or wind protection. Since field tunnels have no sides, wind can get under the covering and lift it off or damage it in areas prone to violent thunderstorms or other severe weather. In places where it snows, gutter-connected tunnels are three-season structures. The plastic has to be either removed or rolled up to the top purlin for the winter, since there is no way for snow to be shed or melted off.

Field tunnels provide a growing environment halfway between the open field and a fully enclosed high tunnel. They will keep the rain off and make things warmer when the sun is shining. Growers also don't have to worry as much about ventilation, since there are no sides and no ends. Their simplicity is the source of their benefits as well as of their drawbacks. No moving parts and no mechanical failures also means no way to overcome serious climate obstacles. These are structures for growers who want something simple to construct and manage – something that gets them into protected culture for the lowest cost per square foot. When the weather is bad or rainy, the only real advantage they offer is that your leaves are staying dry and your crop isn't getting an unplanned irrigation. They improve the weather when it's already good, but you need some decent weather in the first place for them to improve upon.

Simple Hoophouses

Technically, field tunnels are hoophouses, just without sides and ends. But since "hoophouses without sides or ends" is a mouthful, we'll say that open sides and end walls = field tunnel, and sides and end walls = hoophouse.

Adding sides and ends to a field tunnel offers several improvements, including more control over the environment. This makes a big difference on days with marginal weather – for example, when it's overcast and you want to vent minimally and to trap as much solar radiation as possible.

Sides and ends also give you more control over what enters the structure. You can close the house at night to exclude tomato hornworms and other night-flying pests. If control of pests is desired during daylight hours, hoophouses mean much less square footage to cover with insect netting. Birds can be excluded with larger mesh netting – an important consideration on any farm with chickens, or for anyone who has tried to grow berries or other fruits that wild birds like to eat.

The disadvantage of a hoophouse compared with a field house is that it requires a greater investment in building and covering the structure – especially end walls, which usually have to be framed in and likely will need doors. Rollup sides will need to be built; when a structure is completely enclosed, it will overheat very quickly on sunny days.

The extra investment in covering and end walls is negligible in the long run, however, and should not be the deciding factor in whether to build a field tunnel or

Figure 2.1. This freshly planted cucumber hoophouse has netting installed to exclude cucumber beetles.

hoophouse. Decide on the type of structure based on your climate and what you want to grow. If a field tunnel will meet your growing needs, great, save some money on construction. But don't build one when what you really need is a hoophouse. The money you save up front will be less than what you will lose in production from not having the right structure.

Hoophouses are usually covered with flexible plastic because glass or hard plastic will blow the budget, and the general rationale behind them is to get into protected culture inexpensively. Single-bay hoophouses usually have rollup sides, whereas multibay hoophouses may have peak vents or rollup sides. Most smaller hoophouses have rollup sides since peak vents require electricity and increase the cost and complication of building.

In multibay structures, peak vents will help release the heat that increasingly builds up as structures get wider. The larger a structure is, the harder it is to ventilate with edge venting. The ratio of edge to interior goes down the bigger and wider you get, which is why larger structures are more likely to have peak vents. Peak vents will result in a better growing environment with a cooler, drier climate in smaller structures, too, along with added expense.

Fancier Hoophouses and Automation

There are a lot of improvements that can make hoophouses more automated and help them better control the environment. One of the easiest things to automate is water: All you need is a hose and a battery-operated timer.

End wall vents are a fairly simple improvement to increase ventilation if you don't have peak vents, which

Figure 2.2. This freshly planted tomato hoophouse on the author's farm has deer netting installed to exclude chickens (and deer).

Figure 2.3. A small gothic-style hoophouse. The peak at the top helps shed snow to either side. Note the two end wall vents installed side by side at the peak. The more ventilation the better.

most single-bay hoophouses don't. End vents are a poor man's peak vents, as they will help release some of the heat that builds up at the top of a structure. You can add them to any hoophouse by installing louvers as close as possible to the highest point on the end walls.

Even in the absence of electricity, vents can be opened by a rope or chain tied to the rib that opens and closes the louvers. It can be pulled and tied in the open position during the day and released to close at night, though if you do have electricity, it's much more convenient to install a thermostat and a motor to control end vents.

If you are going to install end vents, make them as large as possible, or install multiple vents at each end. There is no way to get enough surface area for ideal ventilation out of end vents. This is why peak vents are better, though they are prohibitively expensive for many small hoophouse projects. But end vents are better than

Figure 2.4. Peak vent and rollup sides in use on the same greenhouse.

nothing when it comes to dumping some of the heat and humidity that build up in the top of a structure.

One easy improvement to make is to set up a remote monitoring system to keep tabs on the temperature; this can help you decide when to ventilate or to cover up a crop. There are many brands of thermometers that can operate remotely in a structure, sending data to a receiver, computer, or smartphone to keep tabs on the temperature. These can be battery-operated in the absence of electricity. Thermostatically controlled end vents can act as a crude temperature indicator. As long as you know what temperature they're set at, you know the structure is cooler than that if they're closed, and warmer if they're open.

Unless you have the skills to do it yourself, it can be a significant cost to run electricity to a hoophouse. If hoophouses are situated far from other buildings that have electricity, it may be very expensive to run power lines to them. For hoophouses *really* far out there, solar may be the only option.

Electrifying a hoophouse does allow you a few simple improvements, including thermostatically controlled end vents and rollup sides, the ability to inflate two layers of plastic covering with a blower fan, and horizontal air flow (HAF) fans. It's always handy to have a few electrical outlets in a hoophouse as well.

Simple Greenhouses

When you put a heater in a hoophouse, you have a simple greenhouse. That being said, there's a big difference between a hoophouse with a heater in it and a high-tech greenhouse. The simplest greenhouse I have ever been in was a single-layer plastic house with a wood-burning

Figure 2.5. One of the author's hoophouses. Note that the plastic covering the large door in the end wall (not the people door) is rolled up and covered with bug netting, to exclude pests and provide increased airflow on hot summer days. The white tube going across the big door carries water to the overhead sprinkler irrigation, which is not in use with a cucumber crop. The duct on the right is pulling in drier outside air to inflate the double layer of plastic.

stove at one end. This is about as low-tech as a greenhouse can be. Though it did allow the grower a much longer season than a hoophouse would, thinking about the disadvantages of such a bare-bones greenhouse illustrates many of the benefits of a more complex greenhouse.

For example, the grower was getting tired of waking up in the middle of the night during the cold part of the year to feed the fire. There was also a big temperature difference between one end of the greenhouse and the other, causing uneven crop growth. Finally, without much ventilation or airflow, the greenhouse built up a lot of condensation in the morning, which led to disease since there was no way to control heat in relation to humidity (see "Heating to Reduce Humidity" in chapter 7).

Fancier Greenhouses

Any of the following additions or modifications to our basic greenhouse with a stove at one end will move a structure up a notch on the scale from low- to high-tech. I'll discuss the options in order of lowest- to highest-tech.

Heat

One thing many growers don't understand is how little difference a hoophouse may make at night. It's not uncommon, especially in the spring and fall, when nighttime temperatures drop quickly after the sun goes down, for the temperature inside a hoophouse to get as low as – or very close to – the outside nighttime temperature. There's only so much you can do to improve the climate without heat.

I have learned this lesson, luckily not the hard way by losing a crop (yet). I used to try to plant tomatoes and other fruiting crops as soon as the nighttime temperatures were out of the 30s Fahrenheit (over 4°C). Inevitably, as soon as the seedlings were transplanted into the hoophouse, the forecast was revised back into the 30s and I would spend my next few nights biting my nails, wondering if the plants were going to freeze.

A layer of floating row cover can be put over the crop at night to protect it in the short term. But the more important lesson to be learned concerning early planting of heat-loving crops in a hoophouse is that they aren't going to grow well until it's warmer. Near-freezing nighttime temperatures are very stressful on a heat-loving plant (not to mention the grower). Even if there is some heat during the day, recovering from nighttime temperature stress will cause the plants to grow slowly.

If growing conditions are poor in the hoophouse, it may be better to hold the plants in the propagation space until weather conditions improve. When weather conditions are bad, the plants are not going to grow very fast anyway. So they are probably better off in a nursery where they are going to continue active growth than in an environment where they are going to stall out until the weather improves.

The other thing to keep in mind about planting heat-loving crops in a hoophouse is that if plants are growing slowly due to adverse weather, poor growing conditions leave them susceptible to disease. You do not want your plants growing in a situation where conditions for disease are more favorable than conditions for the plant. Even if disease doesn't kill the plants outright, they may pick up a problem that will haunt them well into the season, stemming from poor growing conditions at the beginning. It can still be a guessing game to figure out when the weather will make for good growing in a hoophouse. But it is worth planning transplant times around when the plants will begin to thrive, not just survive.

Never use a heater that burns fossil fuels to temporarily heat a hoophouse unless it was designed for that purpose. Plants are very sensitive to ethylene gas and the other by-products of combustion. If you must use a combustion heater, make sure the exhaust is vented out of the structure, unless it is a model that was specifically designed to burn so cleanly that it doesn't need exterior ventilation around plants.

One of the biggest ways people get in trouble with this is when they try to set up an emergency heater in an unheated structure during a cold snap. They can end up killing the crop with the effect of the exhaust gases. Given a high enough concentration of fumes, plants can succumb very quickly.

In almost all fruiting crops, lower concentrations of ethylene gas can cause flowers and fruit to drop off otherwise healthy plants. Higher concentrations can cause leaf yellowing, necrotic spots, and eventual plant death.

Even properly vented heaters can give off dangerous levels of gases if the burner is dirty or not adjusted properly. If you suspect that you have an ethylene problem, an easy way to tell is to let tomatoes be your canary in the coal mine. Tomato plants are very sensitive to ethylene, and one characteristic symptom of exposure is epinasty. *Epinasty* refers to the downward curving of the leaves, which gives the plant a wilted appearance even when it isn't wilting. If you suspect that a heater is venting ethylene into the greenhouse, take some tomato seedlings and put them as close as possible to the exhaust of the heater. If you see symptoms of epinasty, you can conclude that the heater is producing ethylene.

Cracks in the heater or exhaust pipes are a common cause of exhaust gases in the greenhouse. It's not a bad idea to have a technician come to clean and service the heaters every year before you fire up the greenhouse for the first time. He or she can make sure everything is working properly, so your heaters run at maximum efficiency without polluting the greenhouse environment.

Greenhouse Design

The trend in greenhouse design is for taller structures and wider bays, due to the advantages of enclosing a larger air mass. Though it may seem counterintuitive, it's more efficient to heat a larger air mass than a smaller one. It seems like you would want to have as small a greenhouse as possible to save on heating costs, but it's actually more efficient to make gradual changes to a large air mass than to constantly be making changes to a smaller air mass.

Greenhouse designs have progressed over time from narrow, low spans with bows spaced close together to wide, tall spans with bows spaced as far apart as possible. The ideal greenhouse protects a large air mass with as few light-interrupting structural elements as possible. Such greenhouses take more energy to heat in the first place, but once heated, larger air masses fluctuate less and are more efficient to keep at temperature. This is

Figure 2.6. This view of new greenhouse construction built onto an older greenhouse shows the contrast between old and new. You can see how the older house on the right contains a much smaller volume of air. The house on the left is more modern, permitting a taller plant height and more efficient heating with a larger air volume.

why the gutters of new vining-crop greenhouses may be as high as 18 feet (6 m).

Wider bays make it easier to move air within a structure, improving the ease of uniform climate control. Wider bays also have fewer gutters and posts, which means less shade. Since you can always reduce the amount of light coming into a greenhouse, you want to have as much light transmission as possible – and as few impediments as possible – for cloudy weather and low-light times of the year. If you can cover a given area with eight wide bays rather than ten narrower ones, you have covered the same area with fewer gutters, which leads to a higher rate of light transmission. A house with fewer gutters is also cheaper to build, because every gutter you can eliminate reduces the amount of materials, including ground posts, that have to go into the structure.

Greenhouse Coverings

A wide variety of materials can be used to cover a greenhouse, evaluated based on the percentage of light they transmit; the more transmission the better. It's a rule of thumb that a 1 percent reduction in light transmission translates roughly into a 1 percent reduction in yield. Try to get as much light transmission as possible, since you can always reduce the amount later, if you need to, through shading.

When choosing a greenhouse covering, it's also important to consider its strength and shatter resistance. Strength is especially important in areas with anticipated snow load or high winds, and resistance to shattering is crucial in areas prone to hail.

Materials and prices are changing all the time. When it's time to cover a structure, it is worth finding out all the available options. For most applications, you want to look for both a high percentage of light transmission and a high amount of diffusion. If you're going to do it yourself, you'll also want to choose a material that's easy to work with. Coverings fall into three categories: glass, flexible plastic films, and rigid plastic panels.

GLASS

Historically, glass was the only option for greenhouse covering, and many large growers continue to prefer it. It has a lot of benefits, including high light transmission and a long service life. But it also has some drawbacks: It's much more expensive than flexible plastic (the price typically puts it out of range for use in hoophouses), difficult to work with, and fragile.

Whether or not the up-front cost is worthwhile depends on your individual preference and budget. Glass may be expensive in the short term, but not having to replace it every four years may even out the cost over the long run. Many long-season growers also feel that the extra light transmission makes a difference in the winter, especially in northern areas when days are short and light is low.

In addition to lasting longer, another advantage of glass is that it tends to produce a drier environment – for two reasons. First, since glass is applied as a single layer, it doesn't insulate as well as double-layer plastic, which results in more condensation forming on glass. This takes water out of the air. (Whether glass or plastic, most modern gutter-connected greenhouses are designed so that the condensation that does form runs down the covering and into the gutters, instead of dripping onto the crop.)

The other factor leading to a drier climate in glass greenhouses is that they have so many more seams that allow air in and out. Glass greenhouses are covered by many small panes of glass, rather than by a few big pieces of plastic. For this reason, plastic houses tend to be more airtight, while the outside air exchange rate is higher in a glass house. The higher air infiltration and lower insulation value of a glass greenhouse mean that it takes more energy to heat than a double-inflated plastic-covered greenhouse. On the other hand, you may end up needing to heat and vent at the same time more often in a plastic house (see "Heating to Reduce Humidity" in chapter 7).

Glass naturally provides very direct light transmission, but now that we understand the benefits of diffused light, it's possible to buy glass with treatments that diffuse the light.

RIGID PLASTIC

Rigid plastic coverings combine some of the best qualities of glass and flexible plastic. Several materials are available in panes similar to glass. They are not quite as expensive as glass but do provide higher light transmission than most flexible plastic options. Rigid plastic isn't as easy to work with as plastic film, though it is more forgiving and resistant to breaking than glass. It also has a longer service life than plastic films.

Rigid plastic may allow more air infiltration than flexible plastic, due to having seams like glass. But most rigid plastic is at least double layer, so the insulation value is better than glass. Triple-layer rigid plastic is even available for very cold areas where maximizing the insulation value is important.

ACRYLIC

Acrylic panels can be purchased in double- and sometimes even triple-wall versions. Acrylic can offer very good light transmission, as well as better insulation value than glass due to the layers of air in the multiple walls. It's shatter-resistant and does not break down under UV radiation. The downside is that acrylic panels are harder to work with than flexible plastic, and are fairly expensive, though usually less so than glass. Acrylic is also more flammable than other options.

POLYCARBONATE

Like acrylic, polycarbonate is available in multiwall panels, and it has good light transmission. Its insulation value is similarly better than glass given its multiwall properties. It's also impact- and burn-resistant. Polycarbonate is fairly expensive, though.

FLEXIBLE PLASTIC FILMS

Plastic films, mainly polyethylene, are the most commonly used coverings for small- and medium-sized greenhouses due to their affordability and ease of use. Make sure you get greenhouse-grade plastic, which is stabilized to resist breaking down under ultraviolet light so it will last four seasons under greenhouse conditions. Clear plastic films made for construction and other uses are not stabilized. Don't use these; they will break down rapidly under greenhouse use.

One of the advantages of flexible plastic coverings is that they're cheaper than glass and rigid plastic. Flexible plastic is easier to work with and more forgiving of impact than glass. If it does break, small holes in film can easily be patched with repair tape.

Most heated (and some unheated) flexible plastic houses use two layers of film, which are inflated by a blower fan. This provides a layer of insulation in the top of the structure, where heat is most likely to escape. Inflating two layers of flexible plastic provides the efficiency of insulation without the cost of double-wall rigid plastic. Another advantage of flexible plastic coverings is that they diffuse light, which has many benefits (see "Diffusion," on page 23).

A disadvantage of flexible plastic coverings, however, is that they don't transmit as much light as glass. While sacrificing a few percentage points of light transmission for lower cost may not matter, if you're trying to extend the season in northern areas that get barely enough light in winter, a few percentage points might make the difference between not enough light and decent crop growth.

Another disadvantage of flexible plastic coverings is that they usually need to be replaced every four years. Glass lasts much longer. So some growers with glass houses figure that by the time the plastic has been replaced three times over a dozen years, it has cost as much as glass.

If you are going to use a blower to inflate two layers of plastic, it's important to draw your air from the outside of the structure; the interior air is usually more humid due to plant respiration. If you use interior air to inflate, more condensation will form between the layers, further reducing light transmission.

SOLAWRAP

SolaWrap combines some of the best features of double poly and glass; it's been available in Europe for over thirty years but only about five in North America. SolaWrap comprises two layers of plastic that sandwich a layer of bubbles. In fact, it looks like bubble wrap.

Having two layers of plastic held apart by bubbles means you get the benefits of double-inflated poly without having to run a blower fan all the time. SolaWrap is very strong and also very durable, with a ten- to twenty-year service life and no risk of shattering. Though it's more expensive than single-layer plastic films, it has high levels of both light transmission and diffusion.

Figure 2.7. SolaWrap installed on a hoophouse. Note that it looks like bubble wrap, and needs to be installed in strips across the structure.

Plastic Film Treatments

As greenhouse technology has advanced, a number of extra features that can be built right into greenhouse coverings have come onto the market.

ANTI-CONDENSATE TREATMENTS

Anti-condensate treatments are somewhat of a misnomer, because they don't actually prevent condensation. Instead, they reduce the surface tension of the covering so that any water that *does* condense on the plastic does not form droplets. Rather than beading up, the water stays spread out as it slides down the coating toward the lowest point of the covering.

Anti-condensate treatment can be a huge advantage when dealing with two big greenhouse problems: loss of light transmission and water dripping on the crop inside the greenhouse. When water condenses on a greenhouse covering and forms droplets, it blocks some of the sunlight from coming in. The lens-like action of beads of water can act as a reflector, bouncing some percentage of the light that hits the greenhouse back out. When the water stays spread out in a thin film, though, it doesn't intercept as much light coming into the greenhouse.

Keeping condensation on the interior of your greenhouse spread in a thin layer is important because when big droplets form, they will eventually fall on your crop. Since one of the advantages of growing in a greenhouse is the ability to keep the leaves dry, artificial rain falling from the covering is undesirable. For one thing, dripping water may cause watering irregularities. But a more serious problem is that standing water on plant parts will encourage the development of diseases such as botrytis, mildews, and even late blight.

When you're installing anti-condensate-treated covering, make sure the side with the treatment faces the interior of the greenhouse. Plastic is usually clearly marked THIS SIDE DOWN to help you apply it properly.

INFRARED TREATMENTS

Another treatment that can be applied to the interior of the plastic is an infrared (IR) heat-trapping treatment. This increases the greenhouse effect of covered structures by absorbing IR heat that might otherwise pass back out of the greenhouse and radiating it to the crop in the evening hours, reducing some of the need to heat at night when such need is greatest. The treatment is applied during manufacturing, so look for this option when you buy new plastic. As with anti-condensate treatments, make sure to apply plastics with the IR treatment facing the interior of the greenhouse.

COOLING FILM

Some new films designed for very hot growing regions have the ability to reduce the temperature inside the structure, compared with regular plastic, without reducing light levels. This represents an improvement over traditional ways of reducing the temperature under cover, such as applying a white coating (whitewash) to the exterior of the plastic, or using shade cloth.

Cooling plastics are like the opposite of IR treatments, reflecting some of the solar heat without reflecting light. This results in some diffusion as well. However, if you need to reduce light and temperature levels, as in some places with hot, sunny weather, whitewash or shade cloth can be a good option for doing both at the same time.

DIFFUSION

Glass has traditionally been one of the best coverings for a greenhouse. It transmits a very high percentage of the light that hits it, meaning that most of the available light makes it into the greenhouse to fuel plant growth.

It's now known that the *quality* of light – not just the quantity – can have a big effect on growth. There are many benefits of using diffused light, even when it's not as strong as direct sunlight. Light that is diffused, or scattered, by the covering can increase the photosynthetic efficiency of the crops grown under it. This is because light that passes straight into the greenhouse tends to create areas of very high light at the top of the plant canopy, and areas of lower light penetration and shading on the lower leaves. This effect is particularly pronounced in tall, vining crops.

Direct light can create a disparity where the top of the plant is performing a lot of photosynthesis, while the lower leaves aren't. Scattered light, on the other hand, reduces the amount of light hitting the top of the plant, while increasing the amount of light falling on the lower leaves. With diffused light, more leaves photosynthesize with greater efficiency than they do with direct light, resulting in an increase in photosynthesis, especially in tall crops.

Another effect of diffused light is that leaf area tends to increase. Plants compensate for lower light levels by growing bigger and thicker leaves to intercept more light. You will notice that the same variety grown in both an undiffused glass house and a plastic-film-covered house will have smaller, thinner leaves in the glass house.

Diffused light can also lower leaf temperature, especially at the top of the crop, where temperatures are most in danger of getting too high. The head of the plant gets higher levels of light and heat than does the rest of it. Under high-light conditions, the top of the plant may get too hot or too much light, which reduces the amount of photosynthesis. By spreading out the light, diffused coverings reduce the likelihood of too much light or overheating.

The overall effect of diffused light is that it can result in a net increase in photosynthesis over nondiffused coverings. The increased photosynthetic efficiency can more than make up for the amount of light excluded by the covering, and result in increased yields.

Head House

Work space attached to a greenhouse is known as a head house, which is another way to upgrade your structure.

Efficiencies are gained by centralizing all the jobs that go along with growing plants into the same structure. One obvious advantage is that in freezing weather, produce doesn't have to be protected on the way out of the greenhouse to a separate packing area.

A head house may also have space for offices, employee lockers, packaging storage and a packing area, coolers, produce washing stations, a lunchroom, bathrooms, and anything else that might be necessary for the day-to-day operations of the greenhouse. When everything is located in the same building, supplies go into the greenhouse, and trucks pull up and haul the produce away without it ever having to leave the structure.

If you are building a new greenhouse or adding on to one, the easiest way to build a head house is to figure out how much space you need, and – if you live in the northern hemisphere – add that much to the north end of your structure, using the same materials, so that it won't cast a shadow on your production area. This is usually cheaper than building a brick-and-mortar head house, and it will have a seamless connection with the main greenhouse.

The head house can then be customized with separate features from the greenhouse. A wall or plastic curtain is put up between the greenhouse and head house so the climates can be controlled separately. White plastic is usually used to cover the head house so it doesn't get as hot as in the greenhouse.

Whether or not there is a head house on the north end, it may be advantageous to glaze the north wall with opaque insulating material instead of greenhouse glazing. Since in the northern hemisphere the sun shines from the south, this reduces the amount of heat escaping from the north wall without blocking much usable light. Greenhouse builders or covering manufacturers can advise you on what materials are available that have better insulation properties.

Many growers choose to put the people doors or barn doors on their head house, rather than the greenhouse itself, so the head house acts as an "air lock" – an intermediate space between the greenhouse and outside air conditions. When it's freezing outside, this prevents exterior air from going directly into the greenhouse every time someone walks in or out.

CHAPTER THREE

Heating, Cooling, Lighting, and Irrigation

One of the great things about greenhouses and hoophouses is that they are infinitely customizable. The structure itself is really only part of the picture when it comes to the capabilities of protected culture. There are many different ways to modify the basic structure with heating, cooling, lighting, irrigation, and other options that help growers meet their specific needs.

My own farm is a good example. We have three different types of protected culture houses, all of which are very similar structurally, but their features differentiate what they can do.

On the low-tech end, we have a simple hoophouse. It has rollup sides to permit ventilation, and drip tape on a timer to automate irrigation. We built this ourselves and it's not as tall and sturdy as our other structures, so we use it to grow low crops that don't need trellising, like hot peppers that we allow to bush out and grow field-style.

Slightly higher-tech are our "fancy" hoophouses. Though they still don't have any heat, these houses have a few features that make them better for growing tall, trellised crops. We paid a little more for extra-long ground posts, which make the whole house taller. This puts the trusses where we put our trellis wires, 9 feet (3 m) off the ground, which we find to be a good height for our vining crops.

Though it was relatively expensive, we ran electricity over to these houses to help automate some features. Since our houses tend to get too hot in the summer, we installed end wall vents that open and close via a motor hooked to a thermostat. That way we can roll up the sides in the morning, but the vents don't open until things heat up more later in the day. This saves us a lot of time when we're off working elsewhere, and the vents open on their own as the house heats up. It means they always open at the right temperature, and eliminates the possibility of things getting too hot because someone forgot to open them.

The installation of horizontal air flow (HAF) fans that circulate air within the structure, keeping the air fresher and plants drier (see "Active Airflow" on page 35), greatly improves the growing environment. Electrification also allows us to run small blowers to inflate the two layers of plastic covering the house. This gives us an extra layer of insulation on the top of the house, where we lose most of our heat (see "Flexible Plastic Films" on page 21). Electrical outlets are handy for plugging in temperature sensors, power tools, or a speaker to listen to music while we work.

Very similar to our fancy hoophouses is our greenhouse. It has all the features of the fancy hoophouses, plus a unit heater. The addition of heat gives it a degree of versatility that the hoophouses don't have. We use it as a nursery in the winter and spring, and to grow and cure crops during the season.

Which features are chosen often comes down to budget, crops grown, growing area, and philosophy. For example, many growers start out with a hoophouse, realize the limitations of growing without heat, and install a heater or build a heated structure. On the other hand, some growers who could afford to do otherwise stick with hoophouses, preferring to grow without heat as a matter of style. This chapter will help you make sense of the many options that are available to customize a hoophouse or greenhouse.

Heat Sources

The best source of heat depends a great deal on what's available in your area. For many greenhouse growers, the cost of heating a greenhouse is second only to the cost of labor. Choose your fuel source carefully, because it will have a big impact on your bottom line. A lot of greenhouses went out of business in the 1970s during the energy crisis, when spiking energy costs blew many budgets.

Though heating has gotten more efficient since then, technology alone may not be enough to keep growers in business through fuel price fluctuations. As renewable heating options improve, these sources will become a more important piece of the greenhouse heat puzzle, because they displace some of the need for fossil fuels.

The simplest thermostat I ever heard of belonged to a grower who told me he would sleep out in his tomato house in the springtime when the nights were still freezing. If he woke up and his nose was cold, he would turn on the heater.

It may be romantic to sleep out with your tomatoes and take care of them when they need it, but having a heat source that is self-feeding (and thermostatically controlled) is more than just a convenience. Self-controlled heating will even out the highs and lows in temperature caused by a direct heat source, such as a woodstove, as it heats up and cools down. Plants appreciate consistency and will grow better with heat that stays evenly at the right temperature.

Most of the commercially available options for heating greenhouses use self-feeding fuel and self-starting heaters. There are some heat sources that are not self-feeding, like most biomass or wood boilers, which can deliver more consistent temperatures by heating water, which is used to heat the air.

Organic Matter

Cordwood, wood pellets, and biomass can all be used to heat a greenhouse. Biomass can include many types of organic matter; heaters have been designed to handle a wide range of inputs, including agricultural residues, wood by-products, and even grain. Wood and biomass may be competitively priced compared with other fuels, and some growers appreciate sourcing their fuel locally. If there is industry in your area that generates organic matter as a by-product, find out if it can be obtained cheaply to fuel a biomass heater. On the other hand, the size and volume of cordwood and biomass may complicate delivery and feeding. Wood pellets and biomass that can be made a uniformly small size can be used in self-feeding systems with an augur.

Propane, Natural Gas, Heating Oil, and Electricity

Propane and natural gas have become fuels of choice for many growers because they're usually among the most competitively priced options, and are easy to deliver and use. Natural gas tends to be cheaper, but it's not usually available for delivery. So if they have access to a natural gas pipeline, growers usually use that; if not, they use propane.

Electricity and heating oil tend not to be cost-competitive, so they aren't usually used in new greenhouse construction.

Geothermal

Geothermal heat is the hybrid car of the greenhouse world: expensive up front but cheaper in the long run. Two different types are used to heat greenhouses: deep geothermal and ground-loop geothermal.

Deep geothermal involves drilling a very deep hole — similar to a well casing — to access the earth's core heat, the amount of which stays fairly constant year-round. Ground-loop geothermal uses trenches or holes in the ground to access the heat in the earth and soil around the greenhouse and build it up to a usable temperature using heat pumps. With ground-loop geothermal, the amount of available heat fluctuates with the seasons, with less available in the winter and early spring. For this reason it's important to have a backup system capable of heating the entire greenhouse for cold late-winter/early-spring temperatures.

Both types of geothermal heat – though expensive to install – are cheap to run, with multiple units of energy generated per unit spent to run the pumps that bring energy out of the ground. Geothermal is also considered a "green" heat source, meaning it's carbon-neutral.

If run by solar electricity, geothermal could be a viable off-grid heating option. Ground-loop geothermal heat pumps can even be run backward in the summertime, providing inexpensive cooling without any modifications to the heating system. Running the system backward has the additional effect of helping to recharge the heat reserves in the ground that power the geothermal system.

Again, the disadvantage of both types of geothermal systems is a very high up-front cost, though payback times can be relatively short – under five years in some cases. Payback times will be shorter where fuel costs are higher. In areas where there is geothermal heat close to the surface (like hot springs), deep geothermal will be more affordable, since much of the expense in deep geothermal comes from drilling the very deep shaft needed to access the earth's heat in most areas.

Though geothermal is presently one of the rarest heat sources for greenhouses, there are an increasing number of greenhouse projects getting their heat from it. As fossil fuel prices go up over time, and as more geothermal systems are built and prices come down, it will become a more viable source of energy. In some places there may be green energy grants or incentives available for geothermal heating projects.

Heat Delivery

Once you've chosen a source of energy for heating, your next decision is how best to deliver the heat. Determining the heating needs of a structure, and the best way to meet them, can be very complicated. Because the technology is changing all the time, it's worthwhile to get a few different quotes. Companies that sell heating equipment should be able to help you determine the heating needs of your greenhouse.

Vendors should ask for details about your structure, including size, covering, and the weather where you're located. They will also need to know what you want to happen inside the greenhouse; knowing what temperature you want to maintain at the coldest part of the year, depending on what crops you're growing, will help them determine your anticipated maximum heat load, and therefore what options exist for heaters and what delivery methods meet your needs. With quotes in hand from several companies, you can compare both what they think your heating needs are and the cost of meeting them.

Forced Air / Convection Heating

Simply blowing hot air where you need it is a common method of delivering heat in a greenhouse. The fancy name for this is *convection*, or using air circulation to transfer heat. This is the natural method of delivering heat from furnaces, which heat the air directly. Boilers, which by definition heat liquid, can also deliver hot air with the help of a heat exchanger in the system that takes the heat from the liquid and turns it into hot air.

Forced-air heating can be accomplished with one or more unit heaters placed throughout the greenhouse. Since the effectiveness of forced air depends on the hot air reaching the plants, it's important to have HAF fans circulating the air to create an even climate.

When you're purchasing unit heaters, it's important to determine if they are the type that can be vented directly into the greenhouse or not. Plants are very sensitive to the by-products of combustion. Some heaters designed for greenhouses burn so cleanly they don't need to be exhausted, whereas other unit heaters require that the exhaust be vented outside the growing area. The effects of exhaust gases can be so pronounced that a single night of exposure can cause major damage or kill plants. In fact, most of the modern clean-burning unit heaters have an exhaust gas sensor with an override that shuts them down if exhaust gases exceed the range that plants can tolerate.

Where central heat is used, a common way to ensure even heat distribution is to channel the air down

Figure 3.1. A unit heater.

perforated flexible plastic ducts. These tubes, which look like long, skinny plastic bags with holes in them, channel hot air the length of a greenhouse, releasing heat through the perforations. Duct tubes are often placed under tables for bench production, and under or beside crop rows for vining-crop production. Because the heat rises from floor level, it has the opportunity to warm the entire plant on the way up to the roof. Tubes are also sometimes placed overhead where it isn't practical to put them on the ground, though (because heat rises) these don't warm the ground level as well.

Radiant Heating

Radiant heat involves making surfaces hot so that heat radiates off them and rises up through the greenhouse. Circulating hot water through pipes is a very common method of heating larger greenhouses. Distributing pipe throughout the greenhouse also makes

Figure 3.2. This type of unit heater burns so cleanly that it doesn't need to be vented. This simplifies placement, since exhaust gases don't have to be ducted or otherwise vented from the greenhouse.

Figure 3.3. The overhead plastic tube is perforated and distributes heat from a unit heater the length of the greenhouse for a more even climate. This is a way to eliminate the hot and cold spots that can be a problem with unit heaters.

Figure 3.4. This is a fairly typical setup in a modern tomato greenhouse. Double rows end at a concrete slab work area. The slab helps support the ends of the pipe rail used for heating and to guide carts. Pallets of cardboard flats are ready to harvest into.

Figure 3.5. A modern hydroponic tomato greenhouse. The scissors lift used to maintain the heads of the plants is visible to the left of the facing row. You can see the vines snaking around the end of the double row. The scissors lift uses the pipes that distribute heat as tracks to move down the rows.

central heating possible, with one large boiler sending hot water to all corners of the structure. This is usually a more cost-effective option that gives a more even climate for heating a larger greenhouse than siting multiple unit heaters throughout.

Water for radiant heating is usually heated to between 120 and 180°F (49–82°C) in iron pipes, in what's known as a pipe rail heating system, because the pipes can do double duty as the tracks for harvest carts and lifts.

Where pipes do not need to be used as rails and are not in danger of being stepped on, they can be made with fins that increase their surface area and efficiency of heat transfer. This also means that lower water temperatures can be used with finned pipe than with round iron pipe, which may be helpful with ground-loop geothermal and other heat systems that don't make water as hot as a boiler. Pipes with fins are commonly installed around the perimeter of a greenhouse, to maintain the temperature by the walls, where cold can infiltrate. Finned pipe is also sometimes installed under the gutters to help melt snow. The ability to quickly transfer heat can help warm the gutters when snow is building up and prevent greenhouse collapse. Since snow slides into the gutters from the rest of the roof, you can warm the gutters to melt the snow without having to heat the entire greenhouse above normal levels.

Water can also be a way to capture any excess heat generated in the greenhouse if storage tanks are used. For growers who supplement carbon dioxide (see "Carbon Dioxide Augmentation" on page 37), the heat from CO_2 burners is sometimes wasted since augmentation takes place during the day, when less heat is needed. One way to capture the heat from CO_2 augmentation is to pipe the hot water generated by burning during the day into a large tank, called a buffer tank, where it can be stored until needed at night.

Grow Pipes

Some fruiting-crop greenhouses have a heating pipe that runs down the middle of each row of plants. Since most fruiting-crop greenhouse layouts use double rows of plants, this pipe runs between the two rows, usually under the level of the ripening fruit. The small-diameter tube helps make the climate more uniform inside the plant canopy.

The pipe is fed hot water from a flexible tube, so it can be raised and lowered as needed in order to remain just below the lowest fruit on the vine. This warms and dries the interior canopy of the plant where air tends to stagnate, and keeps the fruit warm where it might otherwise be in the shade, which speeds ripening. For low-growing crops like lettuce and greens, the same effect can be achieved by using radiant floor heating.

Grow pipes are expensive to install, though they are yet another investment that can pay for itself in a few years. Introducing heat into the bottom of the plant canopy may reduce botrytis and increase yields without increasing energy consumption. Grow pipes are too small to be the only heat source in the winter months, but they can increase efficiency year-round.

Using a row of LED lights instead of a grow pipe represents a new twist on this idea. Though even more expensive than grow pipes, this setup has the benefit of providing some heating and efficient lighting to a part of the plant that is otherwise usually dark. Though LEDs don't generate very much heat, they throw off enough that they may be able to perform the function of a grow pipe as well as that of supplemental lighting.

Steam

Steam, which was once more commonly used to heat greenhouses, is one alternative to sending hot water through pipes. Because steam is hotter than hot water, it can raise the greenhouse's temperature faster. On the other hand, changes in temperature from steam heating units are more abrupt, making it more difficult to achieve the gradual temperature changes that are desirable for plants. Advances in other heating systems, as well as the high initial and maintenance costs of steam systems, mean that not many new ones are being built anymore.

Conduction Heat

Conduction – whereby heat is transferred directly by touching something hot to what you want to heat, as with heat mats – is the least common method used for greenhouse heating. Where conduction heating is used in protected culture is most often for seed starting, when trays of germinating seedlings are put directly on hot water tubes or electric mats.

Conduction heat is sometimes used on a large scale in greenhouses, where hot-water tubes are buried in the soil or in concrete to heat from the ground up. In very cold climates, these tubes may be laid on top of insulation that is buried to help reduce cold infiltration. Heating the ground or floor can be advantageous, because cold roots can limit plant growth even if air temperatures are optimal.

Conduction heating tubes can make great supplemental heat sources but have limitations as the main heat source of a greenhouse. Although gradual temperature change is desirable in a greenhouse, conduction tubes have to first heat the soil and then the air, making ground heat a *very* slow way to modify the air temperature – too slow to make some of the changes that may be desired for fruiting crops in particular.

Applying root-zone heat also carries the risk of making the root zone too hot. Cranking a ground heat system up all the way in order to melt the snow off the roof of a greenhouse could burn the roots of the plants. You can only make the soil so hot before plant roots begin to suffer, even those of heat-loving crops.

Frost Walls

Frost walls are a simple way to reduce heat loss and keep a structure warmer around the edges. To make a

Figure 3.6. This greenhouse has been retrofitted to accept boards of foam insulation to cover the bottom edge in winter. This provides extra insulation where light transmission isn't important. A frost wall works on the same principle but extends down into the ground.

frost wall, place insulation inside or outside a structure's perimeter along the ground; you might want to extend down into the ground. This increases heating efficiency by reducing cold infiltration through and under the perimeter of the greenhouse – especially important for crops that are grown in soil, since roots around the perimeter of the structure may become excessively cold without one. While it is an extra expense to consider when building, in cold areas a frost wall should pay for itself quickly by increasing heating efficiency and improving crop growth.

Frost walls can be constructed simply by digging a trench along the perimeter of the greenhouse, inserting foam insulation between or around the posts, and backfilling. Professional greenhouse builders may also be able to offer other options depending on the climate.

Using Multiple Heat Sources and Delivery Methods

Many greenhouses employ multiple heat sources and delivery methods at the same time. For example, a greenhouse might use both natural gas and geothermal heat, with delivery through pipe rail and forced air. This allows growers to make the most of the advantages of each type of heat delivery system – plus there's a backup if one of the sources were to become inoperable.

Methods for Cooling

The most basic method of cooling a greenhouse is through passive ventilation, releasing some of the heat that has built up inside. Simply taking advantage of the fact that hot air rises to draw cooler air into the structure is a very effective cooling strategy, and many greenhouses rely only on this principle for cooling.

If passively releasing hot air does not offer enough cooling on its own, you can add exhaust fans to increase the amount of circulation. Active air circulation will increase the number of times the air in the structure can be exchanged per hour. In greenhouses equipped with active ventilation, if venting alone can't keep up with heat buildup, exhaust fans will switch on to provide a second stage of cooling.

Rollup Sides Versus Peak Vents

In a single-bay structure, the ratio of edge to overall surface area is high, so rolling up the sides is an effective way to vent, though heat does still tend to build up in the top of the structure.

As you add bays, the ratio of edge to surface area goes down and rollup sides become less effective. The longer and wider houses get, the more heat will tend to build up in the interior. This is why most single-bay structures have rollup sides and most multibay structures have peak vents.

Peak vents are the best way to passively ventilate a structure. There are a number of designs, but the basic idea is that the vents are situated at the top of the structure, releasing heat where it builds up the most.

Even in structures without rollup sides, peak venting creates a nice climate for plants. Cooler, drier air is drawn up over the plants as hot, humid air escapes. Peak-vented structures tend to be cooler and less humid than ones with rollup sides. The reason you don't see more single-bay structures with peak vents is that they're more expensive to install than rollup sides. But if you're building a single-bay structure and can afford it, consider investing in peak vents.

Single Peak Vents Versus Double Peak Vents

Peak vents can be built with a single vent that opens to one side, or two vents with one opening to each side of the peak of the structure. Double peak vents are also known as butterfly or gull-wing vents, since they look like wings when both of them are open. Double peak vents are more effective than single since they can open up more area for ventilation, but they're also more expensive.

A single peak vent is usually installed so that it opens in the direction opposite to the prevailing wind – in other words, facing out of the wind. This is so that the wind will aid in ventilation as it blows over the top of the house, sucking air *out* of the structure more than if it blew *into* an open vent. Of course winds don't always blow from the same direction, so if you have two peak vents you can choose to keep only the proper vent open. And if it's sunny and calm you can open both peak vents all the way, giving you twice the amount of ventilation.

The performance of peak vents can be enhanced by managing them with an environmental control system (see "Climate Control Systems" on page 36). Especially when the climate controller is connected to a weather station, it can take wind speed and direction into account to determine the most advantageous positioning for single or double peak vents.

Evaporative Cooling

Sometimes called a wet wall or swamp cooler, an evaporative cooler is a device commonly used to actively cool a greenhouse by taking advantage of the cooling properties of the endothermic reaction that occurs when water evaporates.

To make an evaporative cooler, construct one end of the greenhouse as a wet wall: a porous, honeycomb-like material with a lot of surface area, through which there is a constant flow of water. At the opposite end of the structure, place large exhaust fans that pull air into the greenhouse through the wet wall and out the end with the fans. The movement of air through the wet wall causes evaporation, which cools the air and increases the humidity.

Figure 3.7. A portable evaporative swamp cooler being used to cool a greenhouse. Note the green hose going in the bottom of the machine to supply water. Portable units can be handy, but it's much better to install a permanent swamp cooler if this is an important method of cooling for you.

Evaporative coolers are a good option for cooling a greenhouse in hot, dry areas. They only work well in areas that have fairly low humidity, since they depend on the ability of the air to take on moisture to facilitate cooling. If the air is too humid to begin with, swamp coolers won't be very effective. If you need to figure out whether one will work in your area, find out what the humidity level is at the time of year you want to use one, and consult manufacturers. They can tell you the humidity requirements for use, or may know if other growers in your area are able to use a wet wall.

Evaporative coolers also tend not to cool the greenhouse environment evenly. In a structure using evaporative cooling, it's noticeably cooler by the wet wall than it is by the fans on the opposite wall. Air cools down as it's drawn through the wet wall and warms back up as it moves through the house. But uneven cooling is better than no cooling at all.

Portable swamp coolers are available for smaller structures or for temporarily cooling an area. But these are more expensive and less effective than integrating one into the design of the structure. So if you anticipate needing evaporative cooling, build the structure with this included in the design.

High-Pressure Fog

High-pressure fogging can be used to cool down a structure or increase its humidity, or both, via a system of small pipes and nozzles usually installed under the gutters. High-pressure fog is similar to the misting spray used in some grocery stores to keep produce moist. It is integrated into the climate control system, so that when the temperature gets too high or the humidity gets too low, the system comes on. It works by pushing water at high pressure through nozzles, creating a very fine mist that flash-evaporates into the atmosphere. The evaporation cools the air and adds humidity. The advantage of high-pressure fog is that it's so fine, it evaporates almost immediately; you can add humidity without wetting the leaves.

White Ground Cover

While not a cooling method per se, white is the preferred color for greenhouse flooring, partially because of its ability to help avoid overheating in the structure. Black is the standard for plastic in the field, since it absorbs heat and warms the ground around crops. Though it might seem like the same would be desirable in protected culture, it actually contributes to overheating in the summer. Even in mild areas, greenhouse overheating is a common problem. As soon as the sun comes out, the temperature tends to spike, and having black on the ground only makes the problem worse.

Another benefit of white flooring is that light is reflected back up into the crop. As much as 80 percent of the light that hits white ground cover goes back up into the canopy where the plants can use it, making it more effective for increasing light levels than nonreflective ground colors. When the plants are small, much of the light entering a structure ends up striking the floor and bouncing back up to the plants, which can help during overcast spring weather.

If you are growing in soil, use white-on-black plastic covering (the white side faces up; the black side, down). Plastic or landscape fabric that is only white and not completely opaque will still transmit light, allowing weeds to sprout and grow under the plastic.

Active Airflow

HAF fans are one of the most basic improvements that can be made to any structure. They are cheap to install, help even out the temperature in a structure, and keep the leaves drier than passive ventilation alone.

Even if the airflow in your structure seems adequate, microclimates of humid air can develop close to plant leaves as they respire and give off moisture. HAF fans help break up these microclimates and keep plants respiring. Condensation is also less likely to form with moving air, which results in less disease. HAF fans should run constantly whenever there is a crop in the structure.

If there are problems with layers of air developing in a structure, vertical air flow (VAF) fans can help mix them up. VAF fans are specially designed to be suspended from trusses or other hardware in the structure and push air down to break up stagnation.

The Importance of Airflow

The year 2011 was a busy one for us, as we built some of the hoophouses on my farm, plus my son was born in the middle of the summer. So I like to think it's understandable that we never got around to installing the HAF fans until the next year.

At the same time as I was growing hoophouse tomatoes on the farm at home, I was running the hoophouse tomato trial 20 miles (32 km) or so away at the Johnny's Selected Seeds Research Farm in Albion, Maine. Despite the fact that late blight showed up in Albion before it showed up in my area, my tomatoes got late blight, but the tomatoes at the research farm didn't.

The only difference between the two hoophouses that I can think of to explain this is the fact that the Johnny's hoophouse had HAF fans, and mine didn't yet. I installed mine right at the beginning of the next season and I haven't had late blight in the hoophouse since.

Figure 3.8. This greenhouse has fans in the end walls to increase ventilation; the fans are covered with netting to exclude insects. Note the fans in the background that are not covered. A good rule of thumb is that with fine screens, your opening may need to have up to five times the surface area of an uncovered opening to ensure the same amount of airflow.

Climate Control Systems

The simplest way to have a self-regulating climate is to connect your heater and vents to a thermostat such that the heaters turn on automatically at a preset low temperature, and the vents open up at a higher preset temperature. This is a huge improvement over having to facilitate heating and cooling by hand, though it does leave something to be desired.

Say you want to take prevailing winds into the equation, opening only the vents that face out of the wind for maximum venting. Or maybe you want to keep a few vents open at the same time you're heating in order to reduce humidity. These more complex climate control goals are the reason that in fancier greenhouses, temperature control is integrated into overall climate control, which can take into account other factors and grower objectives in addition to the temperature.

Computerized climate control systems have had a huge impact on greenhouse growing, representing one of the biggest advantages protected growers have today over growers in the past. Computers can monitor more factors, and react more quickly, than any individual grower to facilitate more nuanced management goals than simply keeping the temperature within a certain range.

A computerized climate controller is worth the expense; the increased control of a good system can greatly improve crop health and yield. Most controllers include a weather station, which allows the system to anticipate and make changes inside the structure based on what's going on outside. I've talked with growers who have been in the business for generations, who can recall the whole family running around the greenhouse frantically shutting vents by hand in anticipation of a thunderstorm. Now climate controllers can take into account real-time weather data to decide whether to shut the vents or not.

Another important feature is data logging. Computerized controllers make it possible to record climate data from both inside and outside the structure. This is very helpful when making management decisions based on historical weather patterns. Weather data can then be compared with other factors, such as yield data. The beauty of this is that with data you can get as complicated – or stay as simple – as you want. Data becomes more useful over time when you can look back at successive years' worth and see if this year's production is on schedule compared with past years; you can also see if there are correlations with weather patterns or any other factors you may have data for.

Climate control systems can run anywhere from around $1,000 to over $25,000 for more complex ones that will manage a larger area and multiple climate zones. Most climate controllers have control modules that can be manipulated by a computer or smartphone, though the presets will continue to function even if your computer crashes, as long as the power stays on. Look for one that has this ability to run without a live connection to a computer to take some of the potential for digital error out of the equation.

Carbon Dioxide Augmentation

Many growers increase the amount of carbon dioxide (CO_2) in the greenhouse in order to speed up growth rates. CO_2 is one of the necessary components for plant growth, and higher levels will increase yield. It's useful to think of CO_2 as fertilizer applied through the air.

As of this writing, the level of CO_2 in the ambient air is around 400 parts per million and rising. Levels as low as 200 ppm have been measured in rapidly growing greenhouses, even during the summer with maximum ventilation. When a lot of photosynthesis is occurring, plants can use CO_2 more quickly than it can be replenished. When CO_2 levels get this low, growth slows.

Augmentation gives growers the ability to replace what is depleted by growth, and to go above and beyond natural concentrations. Depending on the crop and conditions, CO_2 levels from 800 to 1,000 ppm can increase growth and productivity by up to 50 percent. See individual crop recommendations for the recommended rates of augmentation.

Figure 3.9. The thin, clear plastic pipe visible on the floor between the pipe rails is delivering CO_2 in this heated, soil-based greenhouse. The plastic sheeting between the posts on the right is forming a curtain for a separate climate zone for another crop beyond the tomatoes.

CO_2 is only used when the plant is photosynthesizing, so it's only added during the day. Because it's heavier than air, thin plastic perforated tubes at or close to ground level are often used to deliver it to the crop, where it mixes in with the air around the plants.

There are several methods of introducing CO_2 into the greenhouse. The simplest is to buy liquid CO_2, which can then be metered into the greenhouse automatically by climate control software when it is needed. Currently, liquid CO_2 is off limits for organic growers.

Another option is to produce CO_2 through combustion. CO_2 generators that burn propane or natural gas are available, and can be situated throughout the greenhouse like unit heaters for even distribution. Along with the CO_2, these units discharge heat into the greenhouse, though they aren't large contributors to heating.

In larger greenhouses where multiple unit CO_2 generators are not desirable, you can locate a larger generator centrally and pipe the gas out to where it's needed. This frequently operates in conjunction with a buffer tank to store the heat from CO_2 generation. Since CO_2 is only needed during the day and heating needs are greatest at night, the buffer tanks are used to store the excess heat produced by CO_2 generation during the day, to be used as needed at night. It's also possible to capture the CO_2 produced by the boiler that heats the greenhouse, which is a very ecological option since it makes use of what is otherwise a by-product of heating.

Lighting

Whether you're thinking about building a new structure or trying a new crop in an existing one, you need to know whether there's enough light to grow a crop at a particular time of year. You can consult a DLI (daily light integral) map, or use a meter that can measure PAR (photosynthetically available radiation; see chapter 5). If you have a climate control system, it may have a light sensor that can tell you how much light the structure is receiving.

If you realize there isn't enough light in your area to grow your chosen crop, you have three options. You could grow a different crop that needs less light — like lettuce or greens — instead of a fruiting crop; you could relocate to a sunnier area; or you could install lighting to supplement what's available naturally.

Lighting a greenhouse, regardless of what type of lights are used, is a major expense. Carefully evaluate whether the value of the production enabled by artificial lighting pays for the cost of the lights and electricity. One way to get around having to light the entire production area is to light only the propagation area, and have the propagation period coincide with the dark time of year. By the time plants make it to the production house, light levels will have increased to the point where the crop no longer needs supplemental light.

Light limitations explain why people don't try to grow tomatoes through the winter in the North without supplemental light, regardless of how much heating capacity they have. For year-round production in lower-light areas, one strategy is to grow fruiting crops from the early spring through fall, and quicker crops with lower light and heat needs like the leafy crops through the winter.

Energy Screens, aka Heat Curtains

Energy screens are a very common feature in modern heated greenhouses. Though they were originally designed exclusively for trapping heat, there is now a wide range of screens available that can also modify the amount and quality of light that enters the greenhouse while deployed.

An energy screen works like a blanket that rolls out over the trusses in the top of the structure. It seals off the area above the crop, limiting the space growers must heat to where the crop is, which is important since the highest heating costs are incurred at night when there is no light transmission.

Specialized materials have been developed so that screens can do more than simply block the transmission of heat. If you're interested, it's worth talking to vendors about what's typically used in your area, because having a screen adapted to your specific growing conditions can be a big advantage. For example, if you are growing in a house covered with double poly or some other material that already diffuses the light, you may want an energy screen that is very clear and maximizes light transmission. The advantage is that you may be able to leave your screen deployed during part of the day, further reducing heating costs by using it even when it's light out. For hot climates, energy screens that block some infrared light can be installed, shading and cooling the greenhouse. Many screens are made of materials that conduct humidity but not heat, so that it's possible to vent even when the screen is in use.

Covering the entire greenhouse with a single solid screen isn't usually possible, since the area above the trusses is blocked by structural supports at every arch. Accordingly, most energy screens are sectional, with pieces that cover the area between each arch. Screens are deployed using a system of motorized cables that pull each section across until it meets the section at the next arch. When fully deployed, they form a barrier that separates the air below the trusses from the air above.

Energy screens can reduce heating costs significantly. However, they are very expensive to install, so they aren't practical for growers who aren't paying to heat a house. Though they can pay for themselves within a few years, energy screens represent an instance where it's important not to confuse price and value. Before you look at the cost of installing an energy screen and cross it off the list as too expensive, figure out how long it will take to pay for itself. The vendor may be able to help you with these calculations, based on how much you're heating now and how much you're projected to save. Because after it has paid for itself, it will put money back in your pocket in the form of energy savings.

The best time to install an energy screen is when a greenhouse is built. They can be retrofitted to existing structures, but this is more costly. Many greenhouse manufacturers offer an energy screen as a standard option with a new structure.

The basic features of an energy screen are insulation value, light transmission, and level of light diffusion. For example, some screens prioritize insulation over light transmission or vice versa. If you're considering buying a screen, get quotes from several vendors. They should be able to make a recommendation based on your area, your type of structure, and the crop you're growing or planning to grow.

Shading

Some growers may be able to get double duty out of an energy screen and also use it for shading. The ideal situation is to use the same screen for insulation during the winter and shading during the summer, since typically an energy screen's higher insulation values also block a higher amount of light. Depending on your needs, it's not always possible to get a single energy screen to do both jobs.

Some growers who need a high insulation value in the winter (for example, glass houses in cold areas) may not be able to use the same screen for shading. In these cases, growers who also need shade can either install a second screen for shading, or apply whitewash or shade cloth to the structure during the hottest times of the year. Whether to shade or not is a matter of opinion. Reducing light intensity is one way to minimize damage from excessively hot temperatures. In areas where temperatures frequently climb above 86°F (30°C) for fruiting crops and over 80°F (27°C) for most leafy crops, shading is frequently applied. The advantage of a screen is that you can deploy it only during the hottest parts of the day, whereas whitewash is usually applied at the beginning of the summer and allowed to gradually wash off over the course of the season.

Applying shade cloth inside or over a structure is a lower-cost option for getting the benefits of shading. To turn an entire structure into a shade house, very large pieces of shade cloth are available to cover a tunnel. Some suppliers offer cloth with grommets at the edges, which you can use to tie the cloth to a structure.

As with energy screens, there are a lot of different options with shade cloth, with various percentages of shade and different materials available. One alternative to putting shade cloth over a house or tunnel is to hang it inside from the trusses or other interior structures. A single large piece can be attached to the trusses, or long pieces can be put over the trusses, providing a partially shaded environment.

In addition to reducing the damage from excessive light exposure, shading can produce a larger leaf size. This can be an advantage in the production of basil and other leafy crops, as long as the shade isn't so great that the plants get leggy.

Figure 3.10. A freshly transplanted grafted tomato crop in one of the author's hoophouses. A line of drip tape is placed on either side of the row of plants to keep the bed evenly moist.

Irrigation

Like everything else in soil growing, irrigation is complicated. On a very basic level, this is because different soil types vary widely in porosity and nutrient-holding capacity. Generally speaking, sandy soils are more porous and do not hold nutrients as well, and heavier soils hold more water and fertility. The place to start developing a strategy for watering is with a soil test. Most of the soils in North America have been classified based on their agricultural properties. Looking at a soil map will help you get an idea of the type of soil you're dealing with. This will at least tell you whether your soils are on the porous or the heavy side, which will give you a better idea whether your crop will need more or less water.

The best way to make irrigation easier is to control it using a timer or automated climate control system. At the very least it will be one less thing for you to remember to do. Also, automated systems make pulse irrigating possible by breaking the day's irrigation up into intervals. For example, if you know you want to apply an hour's worth of water per day, you can schedule four 15-minute irrigations. In the absence of rain, and with an optimized climate under protected culture, crops grow fast and need a lot of water. Without some form of automation, irrigation would be a full-time job.

As for the best way to deliver water to the crop, there are a number of options to consider. Irrigation systems

Irrigation Strategies: Feel Versus Precision

There are two basic strategies you can use to decide how to deal with watering your plants: You can either try to be as precise as possible, or do it by feel.

Irrigating by feel can work well for some growers; if you're not used to managing irrigations, however, there is a learning curve. The basic idea is close, careful observation, especially at the beginning of the season when you're getting your irrigation scheme figured out.

Obviously, smaller plants need less water, so you can start the season watering less frequently and increase over time to more frequent and larger waterings. Or you can use what's sometimes called pulse irrigation, watering frequently with smaller amounts of water throughout the day. For example, instead of giving plants an hour of water a day, pulse irrigation might give them four 15-minute waterings. Pulse irrigation can help preserve fruit quality in crops that are susceptible to damage by large fluctuations in soil moisture, like tomatoes (which can crack) or cukes (which can curve). The length and frequency of irrigations can also be used as a crop steering tool (see chapter 7).

When you're learning to irrigate by feel, it's important to take cues both from your plants and from the soil. Of course, wilting is an obvious sign, and something you want to avoid – by the time your leaves have wilted, plant growth has already been interrupted. You also want to make sure between irrigations that water doesn't remain pooled at the emitters. It's important for there to be at least some drying between irrigations to draw oxygen into the soil for plant roots and microorganisms. Over time you will gradually develop an idea of how much water the crop needs, increasing it as the plants grow and the season gets hotter, and decreasing it as the weather cools back down and plant growth slows.

If you're trying to be precise with soil irrigations, rather than just going by feel, you can calculate the flow rate per foot of drip tape and get a pretty good idea how much water your crop is getting based on how much drip tape you have deployed and how long it's turned on per day. An even simpler method is to use one emitter per plant; its flow rate will tell you how much water the plant is getting.

There are various ways to monitor soil moisture, such as tensiometers, watermark sensors, or simply touch. In my experience sensors are not as widely used in protected culture as they are in the field, since protected growing areas are usually small relative to the field, and have fairly uniform soil conditions. But soil moisture meters can be used as a "second opinion" to help with water management.

If you use emitters, you can even install some extras throughout your crop whose water goes directly into bottles to monitor quantity. Some soil growers use a rate of 100 milliliters per plant per irrigation for pulse irrigations. If you want to get really precise with soil irrigation, you can sync your system with a weather station and base irrigation timing on the amount of sunlight received.

are divided into two broad groups: sprinklers, which broadcast water over an area, and drip irrigation, which delivers the water precisely to where it's needed. Drip irrigation is commonly used with both leafy and fruiting crops, whereas sprinklers are almost never used with fruiting crops since their leaves need to remain dry.

Ground-Based Sprinklers

Sprinklers that lie on or are anchored to the ground can be a simple and effective way to water. Even a yard sprinkler can work in a pinch, if you can find one that waters in a pattern that works with the shape of your structure.

We used this method as a Band-Aid before we installed overhead sprinklers in our hoophouses.

The main advantage of ground-based sprinklers is that they can be used temporarily and are more easily moved than overhead sprinklers, which are usually installed in the overhead framework of a structure. This can be an advantage in structures that have frequently changing watering needs, allowing you to not water unplanted areas simply by removing sprinklers.

The drawback with any type of ground-based irrigation, however, is that piping and sprinklers may get in the way of operations on the ground. Lawn and garden sprinklers are also not very precise, which can result in some growing areas getting wetter than others. Another problem with using ground-based sprinklers is that they tend to throw water high in the air, which will ruin the spray pattern if the spray comes in contact with the covering. You may be able to play with water pressure to get the throw of the sprinklers to fit your structure better.

Mini sprinklers designed specifically for irrigation and frost protection are a more precise option for growers who want to keep sprinklers temporary and don't want to install them overhead. Many different designs and heads with different spray patterns are available. Some sprinklers have wide bases to keep them stabilized on the ground; others have spikes that can be jabbed into the ground wherever irrigation is desired. Most sprinklers designed for commercial agricultural use (rather than home and garden) do not throw the water as high, which reduces the amount of contact with the covering and may reduce the amount by which the pattern is thrown off by wind. Commercial-use sprinklers also tend to be modular and highly customizable to a variety of conditions and uses. For instance, you can buy sprinkler heads with smaller spray patterns if you only want to irrigate a part of the structure at a time.

Overhead Sprinklers

I usually prefer overhead irrigation in protected culture, because of the time it takes to move ground-based irrigation around and the possibility of accidentally dragging the hoses over the crop. If sprinkler irrigation is going to be a regular feature of your operation, it is worth considering installing permanent overhead sprinklers. Most of the makers of commercial agricultural sprinklers make versions that can be installed for ground or overhead use. Just make sure you get the right sprinkler heads for the way you intend to water, because the design is different depending on whether the spray will be directed up or down.

The trusses in most structures make a great framework for attaching overhead irrigation. A main header line can be run into the trusses, and split into as many rows of sprinklers as needed. If you foresee a need to irrigate certain blocks of the greenhouse without watering others, shut-off valves can be installed where necessary. Though less precise than other forms of irrigation, like drip, under many circumstances sprinklers may be the best way to water soil-grown greens or other low-lying crops where drip tape is impractical.

Drip Irrigation: Tape

Drip irrigation takes two main forms in protected culture: tape and dripper stakes. Drip tape is used just as it is in the field, with long, flat tubes deployed along the row where water is needed. A header, or supply line, will need to run along one end of the greenhouse, to which each individual tape can be connected. Drip tape is a great option for protected culture because it applies water precisely with very little waste.

Drip tape delivers water right to the soil, with very little opportunity for runoff or evaporation. This is good both for efficient use of water and for keeping the humidity down. Since the tape is pre-punched, it can be rolled out and installed quickly. Under most circumstances, it will require some soil staples to keep it in place. Since empty drip tape tends to heat up in the sun and expand, then contract when it fills with cold water, repeated cycles of expansion and contraction will eventually cause drip tape to snake out of place if not anchored.

With crops that have large root systems, like the vining crops, drip tape works best when placed on both sides of each row, or in as many as three lines in the case

Figure 3.11. An overhead sprinkler used to water hoophouse crops. The water is pumped through the black pipe and out sprinkler heads like these running the length of the house. Note the unusual ventilation style in the background. Multiple strips of plastic are fitted over the hoophouse side-to-side, and the pieces of plastic are spread apart when ventilation is needed.

of double rows (one line of tape down the outside of each row, and one in the middle, between the two rows). With smaller crops, like head lettuce, two rows of plants may be able to share a single line of drip tape.

There are a number of options to consider when choosing drip tape, which is available in various perforation schemes with holes closer together or farther apart, typically corresponding to the planting spacing of the crop. If you really want to be precise, most drip tape is rated for how much water it will deliver over a given distance at a certain pressure. You can calculate how much water the crop is getting based on how much drip tape you have deployed.

There's also pressure-compensated drip tape, whose individual holes don't start dripping until the whole tape builds up to a certain pressure. This can be helpful if your greenhouse is on an incline or dip. Without pressure compensation the plants at the lowest elevation tend to get water before and after the plants at higher elevation, which can make it difficult to deliver a uniform amount of water to the entire crop.

Most drip tape is designed to be installed with its holes facing up to reduce blockages; particulate matter should fall to the bottom of the tape, and clean water will flow out the holes on the top. Most drip tape has THIS SIDE UP printed on the top for this reason.

Drip Irrigation: Stakes

Dripper stakes are the most precise way to deliver drip irrigation, and are usually only used with vining crops. They take longer to install than drip tape, so the labor investment is more worthwhile on a crop that will be in cultivation longer. Since each stake has to be individually installed, the method is less practical for crops with higher planting densities, such as the leafy crops.

Instead of using a tube with pre-punched holes like drip tape, dripper stakes are used with solid tubing. A typical setup uses a pipe of 0.5 to 1 inch (1.25–2.5 cm) to send water down the row, with a hole punched at each plant's location for a supply tube for that plant.

The most common product used to deliver water to the plant is very thin tubing known as spaghetti tubing. This tubing is precut to span the distance between the supply line and the base of the plant. Each dripper stake has a hole at the top where it joins the spaghetti tubing and a sharp point at the bottom end so it can be jabbed into the soil at the base of the plant. Just as with pressure-compensated drip tape, you can get pressure-compensated emitters to ensure that each plant gets the same amount of water.

Dripper stakes also make precision watering easier. Since each plant has its own dedicated water supply, you can closely monitor how much each plant receives, using a technique commonly used in hydroponics to monitor fluid output and quality.

To monitor output, add a few extra stakes to the system, placing them in containers instead of feeding a plant. You can then monitor the water in the containers to see if the correct volume is coming out as expected. That way, if the irrigation system is delivering more or less water than it's supposed to, you can identify and remedy the problem before the plants are affected.

However, using dripper stakes also has its downsides in that they're more expensive and time consuming to install than drip tape. They can also be more prone to plugging problems. Since each plant has only one dedicated emitter, a blocked emitter can cause a wilting plant very quickly. This makes having clean water that is free of particulate matter very important when using emitters.

Dealing with Snow

Dealing with snow starts at the planning stage of building a structure. Hoophouse and greenhouse growers have very different options when it comes to snow, since greenhouse growers can melt it off while hoophouse growers cannot. Snow removal in an unheated structure depends on shedding.

Snow on the Hoophouse

In snowy areas, it's unwise to build gutter-connected hoophouses, and the plastic should also be removed from lightweight, gutter-connected field tunnels prior to the snowy season. When snow builds up in the gutters of an unheated structure, there's no good way to remove it.

If you live in an area where you can expect to get a substantial amount of snow and want to build a hoophouse, you'll need to build a different type of structure than growers who rarely see snow. The shape makes a big difference. Gothic houses with peaked roofs shed snow much better than Quonset designs with gradually curved roofs.

The other big factor that will help a hoophouse make it through the winter is the spacing of the bows. Some hoophouses in low-snow areas are built with bows spaced 6 feet (2 m) apart. This reduces both the number of bows and the cost compared with building with bows every 4 feet (1.2 m). Though it's tempting to use a wider spacing, 4-foot bow spacing is standard in areas that expect to see snow. This will help the structure support more snow before collapsing.

No structure is immune from collapse. Anyone with a hoophouse must stay on top of storm conditions and be ready to remove the snow if it builds up quickly. Light snow usually isn't a problem if the structure is sturdy in the first place – the sun can melt it off later. But growers who live in snowy areas should always have a plan for snow removal before it starts snowing.

A roof rake looks like a very long-handled, lightweight hoe, and can come in handy for removing snow, since it's able to reach high up on a hoophouse and pull the snow off. I wrap mine in a towel to keep it from scratching the covering.

Another option is to remove the plastic over the winter, or roll it up to the peak of the structure, to avoid snow building up in the first place. An added advantage is that when you let it rain and snow in the hoophouse, salts that can build up in the soil over time in covered growing get flushed out.

Snow on the Greenhouse

Though gutters are problematic for snow removal on hoophouses, they're often the best way to remove snow on greenhouses. Many greenhouses in snow-prone areas have a heating pipe installed immediately below each gutter, so that if snow starts building up, the gutter heaters can be turned up high to melt it. The meltwater will flow down the gutter and make space for more snow to slide into the gutter and melt.

This is more efficient than heating the whole greenhouse to melt the snow, because a smaller area (the gutter) can be heated up to a higher temperature, rather than the entire air mass. Using gutters to melt snow also allows you to maintain a more normal temperature for the crop, rather than having to make the whole house abnormally warm.

If you don't have gutter heat and you do have an energy screen, make sure to retract the screen before

Figure 3.12. These field tunnels have been customized for winter growing in Florida by adding sides for cold days, fold-down vents at the peak for hot days, and pipes attached to the gutters to direct water away from the structure.

Figure 3.13. An example of a very simple Quonset-style hoophouse the author built with pipes bought from the hardware store and bent on site. It houses chickens in the winter and basket-weaved peppers and eggplant in the summer.

trying to melt the snow off. Otherwise the screen will do its job and keep the heat down by the plants, preventing the roof from warming up as quickly.

Snow Removal from the Sides of and Between Structures

Whether you have a hoophouse or a greenhouse, if you live in a snowy area you may need to make plans for removing snow from around your structures. In particular, snow tends to build up quickly along the sides of a tunnel as it slides off. This creates two problems. When the piles build up along the sides, they may interfere with the smooth shedding of snow off the roof. Second, the piles can get so high and heavy that they start crushing the sides of the structure. This might not seem like a big problem, but tunnels are engineered to support weight from above, not from the sides. Structures have been known to get crushed from the side when snow piles get too big.

If you have multiple tunnels next to each other, and they are close enough that the shedding snow piles combine, the piles will build up more quickly and may need to

Figure 3.14. Passive heat from the sun causes snow to slowly melt off the author's greenhouse. Note how snow buildup on the ground is interfering with the snow sliding off. Piles will need to be removed from around houses if they prevent the snow on the house from sliding off.

be removed more frequently. Have a plow or snowblower on hand to move the snow, and a plan for where to put it.

Cutting the Plastic

There is one last-ditch method you can use to save a structure in the event that snow is building up faster than it can be removed, or if heaters fail during a snowstorm. If you think that collapse is imminent, slitting the plastic from inside can save the structure at the expense of the covering. Snow and plastic falling into the greenhouse may cause damage and will most likely kill any existing crops. But it's a better option than crushing the structure.

CHAPTER FOUR

The Economics and Efficiencies of Protected Growing

The costs associated with growing in protected culture are higher than those for growing in the field. It's not uncommon for high-tech glass greenhouses to cost a million dollars or more an acre to build. But even if you aren't building a high-tech glass greenhouse, costs will be high, so make sure the yield justifies the expense.

Dutch greenhouses are relentlessly efficient, which is how Dutch growers get the greatest return on their big investment in protected growing. They know that it's important for greenhouses to have not only a high overall yield, but also a high yield in proportion to the resources invested.

Consider plant spacing. Since there is only a finite amount of space in a given structure, the Dutch cram in as many plants as possible, resulting in plant densities much higher than are found in field production. This makes the most of the structure itself, and also of resources like heat. Dutch greenhouse pepper growing represents an extreme, where stems may be 6 to 7 inches (15–18 cm) apart in double rows that are 2 feet (60 cm) apart. This creates a veritable hedge of pepper plants that goes right up to the line of overcrowding without crossing it.

Determining the return on investment involves taking all expenses into consideration when figuring how much it costs you to produce a crop in a certain space. Beyond construction, it's important to factor in costs for heating, labor, fertilizer, beneficial insects, tools, supplies, seeds, and anything else that goes into production of a crop. Keep in mind that some of these costs will be different for different crops. A good protected culture mind-set would be to think about the cost per square foot or meter just to run your greenhouse through a particular season. This is useful to calculate because the crop will have to return above this level to make a profit.

The calculations from here on out are where it really gets tricky, because some of them will be speculative. How do you know how much of your crop will be sold versus composted due to poor demand? Vegetable farms rarely sell more than 80 percent of what they produce. And how do you know what prices will be like before the season? Contracts may take some of the guessing out of the equation. But pests, diseases, bad weather, and labor issues will keep it interesting.

The bottom line is, do the math and figure out how expensive your greenhouse space is before you decide what to grow in it. Make sure whatever you grow in it covers those costs and that you make a profit on top of that. As for what to grow, I hope to point you in the right direction by limiting the scope of this book to eight of the crops that are most likely to be profitable for protected culture vegetable growers.

Costs and revenues vary widely depending on many factors: size of operation, type of greenhouse, location, and type of crop, to name a few. One constant is that labor tends to be the biggest expense in most operations. Even in heated greenhouses, the cost of energy is usually second only to labor.

The most important ratio to consider is that of costs to income. A very high-value product might justify high expenses. A good rule of thumb to remember is that on average, producers get paid about half the retail price of their vegetables in the grocery store. Bigger growers tend to be more efficient, enjoying larger economies of scale and having others to do the selling for them. And direct-market growers need higher prices to justify the extra labor of selling direct.

Record Keeping

The first thing you need to do in order to determine if your structure is paying for itself is to keep records. If you don't have records, you can't assess how well you're doing. And if you don't know how well you're doing this season, there's no way to know how much of an impact changes have. A lot of computerized climate control systems automatically make records of the conditions they are controlling. But even if you don't have that technology, if you're measuring or controlling something, you should be documenting it.

Weather

If you don't keep track of conditions both inside and outside your structure, it's impossible to know the relationship between the weather, good or bad, and your production from year to year. The longer you grow in the same place, the more useful this will become.

Developing your own baseline weather, yield, and other data over time – whether through an automatic computerized system or otherwise – will help you troubleshoot production problems. If you know what the conditions are usually like, you can correlate a dip in production to cloudy weather causing abnormally low light levels, or a heat wave causing stressfully high temperatures. And if production problems can't be correlated to anything outside the structure, that can tell you it's time to go looking for problems inside.

Harvest and Income

Keeping records of your productivity can help you determine how efficient you were in a given season by relating production volume to inputs, including fertility, heat, and labor. Harvesting the same yield year after year can be a good thing if you can accomplish it with fewer and fewer inputs over time. The only way you'll be able to know how well you're doing is to record what goes into your crop and what you get out of it, including not only the amount you harvest but the quality of production. How much of your crop is top-quality, how much is lower-quality or rejected, and how much goes unsold?

If you don't already have accounting software on your computer, it can be useful to help keep track of price per pound, who buys what, and whether there is more demand for your food than you are able to supply (and from which customers), so you can evaluate the strength of the market and plan for the following year. In addition, food safety protocols that are often required for wholesaling may demand harvest records.

Expenses and Return on Investment

Most people do this for business reasons anyway, but if you keep track of all the inputs into a crop, and compare that with how much you produce, you can really look at the return on investment (ROI). This allows you to figure out how much you earn compared with how much you spent, as well as which crops pay for themselves and which don't. Otherwise there is no way to tell how worthwhile a crop is just by looking at it. You have to crunch the numbers to determine which crops are profitable for you. There are lots of programs that will help you analyze the records you keep.

Labor Tracking and Productivity

Many successful growers keep track of worker productivity. Even if you only have one or two workers, including yourself, it can be very useful to know how long it takes to do tasks around the greenhouse. How will you know how many people to hire for a job if you don't know how long it takes to accomplish – especially in a greenhouse, where the same jobs need to be done repeatedly over a long season? Since labor represents the biggest cost for many growers, making sure you have people doing jobs they are good at can save a lot of money.

Since so much hand labor is involved in protected growing – as opposed to tractors and combines – one of the foundations of the Dutch system is efficiency. Sometimes it's hard to measure and compare the speed of one task versus another, but even if you don't time yourself *every* time you do something, it's worth doing so once or twice to find out the quickest way of doing things, or if a tool designed for efficiency really is faster, and how long it will take to pay off and justify the tool's cost.

Labor-tracking programs designed for greenhouse use do exist, though most of these are probably too expensive for operations without a lot of employees. Apps and programs for labor tracking not specifically designed for greenhouse use are a cheaper option, and may work just as well for smaller farms that don't have a lot of employees to track.

Saving Money Versus Making Money

Growers tend to be a frugal bunch; otherwise they wouldn't be in business. That said, there are times when trying to save money in the short term can cost you in the long run.

Reusing Supplies

A tomato grower who wanted to reuse all his supplies once asked me the best way to clean everything; he wanted to sterilize everything from the tomahooks to the twine, trellis clips, and landscape fabric (see appendix C). I said it was fine to sterilize the tomahooks because they are metal, which can be safely sterilized, but I advised him to compost or throw out the rest. Porous materials such as twine and small items like clips have a lot of surface area where pathogens can live until the next season. Attempts to sanitize them may fail if air bubbles or dirt prevent cleaning materials from coming in contact with pathogens. One small infection can lead to a pathogen getting out of control.

If your aim is to save money, and you account for all the labor of taking off clips, re-winding string, and sanitizing everything, it's probably a wash when compared with the cost of buying new materials. If your aim is to be green by not generating trash, there are now biodegradable, compostable versions of all those supplies. It really saves time and speeds cleanup to cut entire plants down, roll them up, and compost them without having to remove trellis clips and strings.

Another good reason not to reuse greenhouse supplies is that those items will eventually fail from physical stress and ultraviolet degradation. The only way to find out if you've reached the point of failure is the hard way.

Field Versus Greenhouse Varieties

Having worked in the seed industry, I've seen confusion over the difference between cost and value many times, especially in protected culture, where specially bred varieties cost much more than the same species bred for the field. I've visited many growers who tried to save a few bucks at the beginning of the season by using field varieties, only to lose out at season's end.

Leaf-mold-resistant tomatoes are one of the best examples I can think of when it comes to the relationship between higher price and value. Leaf mold is the kind of disease where once you get it in a structure, you usually have it every year after that – it's almost impossible to get rid of the spores. If you get it once, it's probably going to be a recurring problem. Since the disease is nearly exclusive to protected culture, most field varieties have not been bred to have resistance to it; leaf mold resistance is usually only found in varieties bred for protected growing, whose seeds come at a higher price point (see "Why Greenhouse Seed Is So Expensive" in chapter 6). Money is often tight at the beginning of the season, so I understand the desire to cut costs by planting field varieties in a greenhouse or hoophouse. But I've also seen so many tomato crops collapse at the end of the season due to leaf mold infestation – crops that might otherwise have stayed productive if they'd had resistance to the disease.

Leaf-mold-resistant varieties are completely resistant to the strains that we currently have in North America. I'm sure the pathogen will mutate and that will change at some point. But right now the resistances work really well. By the end of the season in my own hoophouse, the heirloom varieties (with no resistance) are completely covered with leaf mold, while the greenhouse varieties are clean. So leaf mold resistance can be the difference between your crop going down early to disease or remaining productive until the end of the season.

As a commercial grower it's particularly important for your crop to make it through the whole season. One way to look at it is that the first five months of a six-month season covers your expenses – the seed, the heat, the labor, the supplies – and month number six pays *you*. If your plants don't make it past the end of month five, you don't get paid.

To put it another way, if more expensive seed yields one more tomato (or one more marketable head of lettuce, or whatever you're growing) than a cheaper variety does, it has probably paid for itself right there. Anything beyond that is income you wouldn't have had if you'd used a less productive variety.

Especially if you're building a hoophouse or greenhouse for the first time and plan on growing your favorite field varieties, make sure to try some dedicated greenhouse varieties as well. A lot of work goes into breeding varieties that have advantages in protected culture. See the individual crop sections for the desirable traits for each species. Pick the best variety based on your conditions.

This doesn't just apply to seeds; it's worth remembering that in most cases you get what you pay for. Keep in mind that even though something costs more, it might be a better value than something that costs less.

Basic Principles for Increased Profitability

Though in some ways broad generalizations, the following basic principles and strategies can increase the profitability of protected growing.

Space Crops as Tightly as Possible

In open-field agriculture, it is sometimes advantageous to give plants a lot of space. For example, many growers only plant field tomatoes every other row, to promote airflow between the rows and minimize foliar diseases. But in protected culture, space is limited by the size of the structure, so you want to cram in as many plants as possible. The goal is to plant as densely as plant size, light requirements, and airflow will allow. You only have a limited amount of covered growing space, so make the most of it. Take advantage of the fact that your plants are not getting rained on to plant them at densities that would result in foliar disease in the field.

For example, you can pack tomato plants into a greenhouse much more tightly than you can in the field, though the plants will cast shadows on one another. This is because the top of the plant will receive plenty of light, even though the bottom leaves will be in the shade.

Under sunny conditions, it's easy for a vining plant to reach its photosynthetic maximum over the course of a day. When a plant receives more light than it can use, shade on its lower leaves will not reduce its overall productivity. Therefore you can increase plant density, which in effect gives the extra light falling on one plant to the one next to it. You just have to ensure enough airflow to keep the plants respiring well and free of disease. If you crowd fruiting crops too much, the plants will let you know. Fruit size will go down when a plant isn't getting enough light. See individual crop sections for common spacings for each crop.

Match Heat Levels to Light Levels

There is no point in growing the crop "fast" at a high temperature if there isn't a lot of light to fuel growth. The high end of a crop's temperature range should correspond to high light levels. In other words, don't grow a crop at summertime temperatures with winter light levels. If you grow the crop at a high temperature when light levels are low, you will waste your crop's energy with a high rate of respiration without enough energy coming in through photosynthesis. This will result in low yields and weak growth.

Figure 4.1. This farm maximizes the amount of organic greens that can be planted in a tunnel by only having one pathway—though this does limit the growers to harvesting by starting at one end or the other.

Transmit as Much Light as Possible

All greenhouse coverings block some light. Even the clearest glass will exclude a small amount of the light that hits the greenhouse. Since light is what fuels photosynthesis, the more you can allow into the greenhouse, the better production will be. A rough rule of thumb is that 1 percent less light = 1 percent less yield. Beyond picking a covering that transmits a lot of light in the first place, keeping light levels high means maintaining the covering properly. Replace plastic when it reaches the end of its life and becomes cloudy. For glass or plastic in areas with little rain, clean the dust off the structure on a regular basis.

Maximize the Efficiency of Light That Enters Your Structure

First, minimize the amount of things blocking light in your greenhouse. Don't hang anything from the trusses that doesn't have to be there.

Second, use white ground cover to get more out of the light that is already in your greenhouse. Most of the light that hits a white floor bounces back up into the crop, where it can be used for photosynthesis. A white floor will not only increase the amount of light for photosynthesis but also decrease the amount of light that turns into heat, which will help with temperature regulation.

Pay Attention to Light Quality

Diffused light tends to create a more even climate and result in a higher yield than direct light. When light is diffused, it is distributed more evenly over the foliage, creating fewer hot spots and shady areas. This is more pleasant for people in the greenhouse as well as for the plants.

This more even distribution of diffused light is especially beneficial for tall, trellised crops. With direct light, the highest parts of the plant tend to get the hottest, which can cause overheating at the head of the plant. In addition to being less likely to cause overheating, diffusion allows more light to filter down to the lower leaves. With direct light, the lower leaves of a tall plant contribute very little to photosynthesis due to being in the shade. Diffusion spreads light more evenly throughout the canopy, increasing photosynthesis in the lower leaves (see "Diffusion" on page 23).

Diffused light can also reduce or eliminate shadows inside the structure, which will help the crop grow more evenly. One way to evaluate the level of diffusion inside a structure is to look for shadows. If the shadows from trusses and other things inside the greenhouse are crisp, the light isn't very diffused. With very diffused light, shadows from above may not be visible at all.

Consult some covering manufacturers to see what they recommend based on what's available for your area and crop. Finding the covering that transmits as much light as possible and is as diffused as possible usually leads to the best growing environment.

Maximize Each Plant

Protected culture is high-investment, high-return horticulture. One of the ways to maximize yields and extend into a long season is by investing in plant maintenance. Shaping new growth and removing undesirable older growth through actions such as leaf and cluster pruning, as well as topping, will improve plant growth and longevity. For fruiting crops, graft if appropriate to increase vigor, yield, and disease resistance (see chapter 8). Make sure each plant gets what it needs in terms of light, fertility, and maintenance.

Lowering and leaning a vining crop enables you to grow a plant of unlimited length in the finite space of a greenhouse or hoophouse. With it you can grow vines longer than your structure is tall, which allows you to grow plants for as long as you can keep them healthy. If labor is short or a high-involvement management style is not desired, using determinate/bush varieties (instead of tall vining greenhouse varieties), and growing more like in the field can be done instead.

Don't Save Money at the Expense of the Crop

Saving money is great, but it's hard to cut short-term costs – like reducing heating in the early spring, which can be tempting – without reducing long-term income. Corners can be cut, but if they impact the quality of the crop they will impact your bottom line. Decide what level of any given input makes economic sense for your production, and maintain that level unless it's absolutely necessary to cut.

A grower once told me that he wanted to run his greenhouse using less labor, and asked me if I thought it made sense to plant single rows of tomatoes instead of the double rows that most greenhouse growers use. His logic was that by lowering plant density he could reduce the amount of labor he would need. My advice was to compare his expected income against the cost of hiring enough people to maintain a double-row crop. If the expected income would be high enough to justify the labor, go ahead and plant the double rows of tomatoes. If the income wasn't high enough to justify a maximized planting density, though, he should find another crop or get out of the business! If tomatoes returned enough income, they would actually be more efficient planted at higher densities, since he would be heating more plants per square meter than with a low-density planting, saving on heating costs.

Ensure the Best Possible Climate

Even the fanciest greenhouse may not always ensure the perfect climate. But it's important to learn what the "perfect climate" is for a given crop, and always try to get as close as possible to that ideal. Provided that your greenhouse's heating system is sized adequately for your region, it can actually be harder to keep a greenhouse cool enough in the summer than to keep it warm enough in the winter. If you're building a greenhouse in an area that's new to you, discuss heating and cooling needs with local builders to make sure that the climate range for the crop you want to grow isn't unreasonable.

Manage Irrigation for More than Just Watering

One of the biggest advantages of protected culture is the ability to keep rain off the crop. Not only is foliar disease pressure lower when the leaves are dry, but irrigation can be planned and timed with a specific crop's best interests in mind. Through careful irrigation management, you can ensure that your plants never experience water stress from too little or too much at one time.

Though being able to water just the right amount is an advantage for all crops, it's a particular advantage when it comes to preventing splits in tomatoes. Since the skin of the fruit loses elasticity as it turns from green to its ripe color, field growers have to keep an eye on the weather and pick proactively if a storm threatens, to prevent cracking. But protected growers can grow even crack-prone varieties like heirlooms by dividing the total amount of water the crop needs on a given day into multiple smaller waterings. Preventing splits is one of the secrets to making money and having a good amount of salable fruit on notoriously difficult-to-grow heirlooms. Growing in protected culture is one of the best ways to do that.

Use Environmental Control to Influence Rate of Growth and Type of Growth

Protected culture growers can influence both the speed and the type of growth in ways not possible in field agriculture. The average temperature over the course of twenty-four hours determines the speed of growth, with a higher average temperature correlating to faster growth. With fruiting crops in particular, a wider or narrower day/night temperature differential can influence where the plant puts its energy. A wider day/night differential will make the plant more generative (more invested in fruit production), while a flatter temperature profile will make the plant more vegetative (invested more in the rest of the plant). For a full treatment of these topics, see chapter 5.

Evaluate Productivity by Yield Per Square Foot or Meter

It's worthwhile to evaluate yield per plant as well as per square meter. However, the reason protected culture growers use yield per area is that it's a better way to evaluate how well you're making use of the space, not just the productivity of each plant. For example, you could have widely spaced plants with a high per-plant yield and still have a low yield based on the area you're growing in.

It's desirable to have more plants with a slightly lower yield per plant than to have a few high-yielding plants. This is not a strategy that's completely foreign to the field, as many field growers will plant a crop tightly for a lower yield per plant with a higher yield per acre.

Most of the available greenhouse yield data is in terms of yield per square meter, written as pounds or more likely kilograms/m^2, since so much of the greenhouse research comes from Europe. Metric calculations can be easier to consider, as a square meter offers a much larger area than a square foot, which will even out variations (1 square meter equals approximately 9 square feet). But if you prefer to think in terms of square feet, then by all means do so.

Sell Small-Fruited Varieties by the Piece, Not the Pound

Smaller-fruited varieties never have as much fruit on the plant at a time, or yield as much by weight, as larger-fruited ones. To illustrate this point, take all the fruit off a beefsteak plant and a grape tomato plant, and make a pile of each; you will quickly see that the beefsteak pile is larger. The same goes for small cucumbers, peppers, and eggplant. Yield tends to be roughly proportional to the size of the fruit, with larger varieties having a larger yield.

Figure 4.2. Sell small-fruited varieties, like these Artisan cherry tomatoes, by the pint, not the pound, to get a better price to make up for their lower yield and higher cost of production.

One way many growers maximize the yield of small-fruited varieties is by increasing the planting density, especially since many small-fruited varieties also tend to have smaller leaves and more open plants. But yield will still be lower than the same area planted to a large variety, even with a denser planting.

The way to grow low-yielding, small-fruited cultivars and still make money is by selling the produce by the piece, in a container, rather than selling loose by weight. People will pay prices for produce by the piece that they would never pay by the pound.

The same principle can be used for cut greens and salad mix. Instead of marketing your salad mix at $16 a pound, sell it at $4 for a 4-ounce (113 g) bag. That will sound much more reasonable to customers, and cover your extra costs for cutting, washing, and bagging the greens for them.

If you grow a variety that yields half of what another one does, you need to get more money to justify devoting the space to it. It's simply a matter of presenting produce with higher costs of production to customers in a way that they will accept paying a premium, instead of saying, "No way am I paying $5 a pound for cucumbers." Relative to a large-fruited cucumber price per pound, that does sound outrageous. But it's not far-fetched to get $5 a pound (450 g) for cucumbers; we've happily charged $4 for a 12-ounce (340 g) bag of snack cucumbers, or roughly $5 per pound.

The cukes that allow us to charge such a high price are cocktail cucumbers, which at about 4 inches (10 cm) long are the smallest, lowest-yielding type of cucumber. We pack them seven or eight to a bag that weighs approximately 0.75 pound (340 g) and sells for $4. So if you do the math, even allowing for some overpack, we're getting $5 a pound for our cocktail cukes, which sounds expensive compared with the usual selling price of $1 to $2 a pound for most cucumbers, depending on season and variety. However, it's not so expensive when you consider that cocktail cucumbers probably yield half the weight of a large-fruited cultivar, and are much more labor-intensive to harvest. Whereas large cucumbers may weigh anywhere from 0.5 pound to almost 1 pound (225–450 g), I may have to pick a dozen cocktail cukes to equal the same weight. So it takes much more labor to yield the same weight of small cucumbers. That $5 per pound that customers are paying for high-quality specialty produce is the cost of production.

I can't say that growing cocktail cukes is necessarily a way to make more money on cucumbers; it really depends on your market. You will have to crunch the numbers on return on investment based on your own crop costs and prices. But one thing selling cocktail cukes does for my farm is differentiate us from all the other growers who just grow long European cukes or slicers. The cocktail cukes really are a different eating experience, and our market has seen a strong trend for vegetables that can be snacked on and eaten out of hand. We have some customers who prefer these types and won't buy other cukes when we're out of cocktails. What this says to me is that I am drawing customers because I grow something other growers don't have. So I want to make sure I'm making money on the produce in the first place.

One of the biggest values of growing cocktail cukes for me is that they get people in my stand who might not otherwise come in. And hopefully while they're there, they will buy a tomato and some salad mix from me, instead of from the other stands that don't have cocktail cukes. In this way, you can use anything you have that sets you apart from other vendors to draw customers into your stand.

De-Commodify Your Produce

Gasoline is an excellent example of a commodity. Most people don't care what brand it is or where it came from as long as it's available to buy when their tank is running low. Industrial produce is much the same. It doesn't depend on being the freshest, most sustainably produced, tastiest, most nutritious, or a hundred other superlatives that might make you want to buy it. Industrial food's virtues are similar to those of gasoline. It's widely available regardless of the season, and relatively cheap.

Figure 4.3. Grow smaller-framed varieties to increase the number of lettuce plants you can fit in a bed.

Greenhouse growers – even big ones – are trying to de-commodify their produce, which is why there are so many new types of greenhouse produce available in the grocery store, in so many new types of packaging. Greenhouse growers have succeeded in taking much of the market share in the grocery store away from industrial field growers by having better quality. Now that they have taken over the market, they are trying to set themselves apart from one another.

Chances are you won't be able to match the year-round availability and economy of scale that leads to industrial food's ubiquity and affordability. But when the system brings us everything all the time, that usually means produce has been in transport for at least five days when shipped domestically, and probably more like ten days or more for international shipments. Is your produce fresher than that? Organic? Grown with sustainable practices? Pesticide-free? Grown for flavor, not for transport? Then those are your selling points! Think of all the things that make your vegetables great and set them apart from other produce. Then figure out how to communicate that to your customer.

Even if you direct-market at farmers markets, signs and labels can help communicate your message. Because even at a very small market, you can't always have an in-depth conversation with everyone and tell them everything that makes your produce great. If your produce is packaged in any way – even just a produce bag

Figure 4.4. This hydroponic deep water culture (DWC) greenhouse takes dense planting to an extreme by packing in the plants and minimizing the number of pathways. Since the plants are floating, workers can remove them for harvest instead of needing to move into the crop.

that says CERTIFIED ORGANIC — it can help remind your customers why they bought it in the first place.

It's a good farmers market rule that if you get asked a question enough times, it's time to make a sign, and easier than answering the same question over and over again. And if something is confusing your customers, some will just wander out of your stand without asking. If your produce is delicious, tell them so. If people are interested in a particular production method, tell them how it's grown. If they are interested in local food, let them know where you're located. You know what makes your produce special. Make sure all your potential customers do, too.

CHAPTER FIVE

Protected Culture Plant Basics

At the risk of oversimplifying, plants do two basic things: photosynthesize and respire. Photosynthesis is the process whereby plants use the energy in sunlight to turn carbon dioxide, nutrients, and water into carbohydrates, giving off oxygen and water as by-products. Respiration is the process through which those carbohydrates are converted back into energy, which fuels the plant's functions. Though photosynthesis depends on light, respiration takes place constantly, its rate determined by the temperature. These two processes occur at different speeds depending on different factors. It is the job of the greenhouse grower to try to harmonize the speed and amount of both photosynthesis and respiration to optimize plant growth.

For an example of how *not* to harmonize photosynthesis and respiration, let's consider greenhouse growers growing their plants at a high temperature on a cloudy day. In this low-light scenario, the plants aren't generating very much energy through photosynthesis, though they're expending a lot of energy due to the high rate of respiration brought on by the high temperature. This is the protected culture equivalent of mashing the gas pedal to the floor in the middle of a desert without a gas station in sight.

Balancing photosynthesis and respiration involves reacting to ever-changing environmental conditions by adjusting conditions in the greenhouse. If it were simple, the rest of this book wouldn't be dedicated to helping you understand how to use your greenhouse or hoophouse as a tool to create the conditions that will maximize plant growth.

Photosynthesis, Respiration, and Transpiration

How well these three basic plant functions are working means a great deal for how well your plants will grow, and it's largely dependent on the environmental conditions you control with protected culture. Many of the environmental recommendations found in this book are based on optimizing plant photosynthesis, respiration, and transpiration.

Photosynthesis

As noted above, photosynthesis is the basis of plant growth, whereby carbon dioxide from the air and water and nutrients from the roots react with energy from the sun to produce assimilates and release oxygen. Assimilates can include sugars, carbohydrates, proteins, or starches. Unless we are talking about one in particular, it's simplest to refer to the various products of photosynthesis generally as assimilates.

To describe the limiting effects of light, water, and carbon dioxide on photosynthesis, a barrel made out of staves makes a useful analogy. Each of these three factors corresponds to a third of the staves. If all the staves go all the way to the top of the barrel, it can be filled all the way up, representing optimal photosynthesis. But if even one of the staves is shorter than the rest, that one short stave becomes the limiting factor. You can't fill the barrel any higher than the lowest stave. Since photosynthesis depends on the supply of light, water, and carbon dioxide, a limitation on any one of these factors represents the short stave of the barrel. This is why we want to transmit as much light as possible and always meet the plant's water needs. It's also why carbon dioxide supplementation is so common in commercial greenhouses.

In greenhouses that are kept closed during cold times of the year, carbon dioxide levels can easily be depleted so that they become the limiting factor for growth. It may help to think of carbon dioxide as fertilizer that can be applied through the air. Nutrients in the soil around plants become depleted and need to be replaced – and it's the same with CO_2 in the atmosphere around plants. Even during warm times of the year, when ventilation is wide open, augmenting CO_2 levels above the ambient level will increase plant growth.

Levels of up to 1,000 ppm have been shown to pay for themselves with increased growth. Photosynthesis only takes place in the presence of light, so augmenting carbon dioxide doesn't help when it's dark. Starting enrichment shortly before sunrise and stopping shortly before sunset is a common strategy.

Respiration

To recap, respiration is the process whereby plants transform assimilates from photosynthesis back into energy that can be used for basic life functions (metabolic processes) and growth. Respiration involves combining assimilates with oxygen, and the subsequent release of energy, water, and carbon dioxide.

For plant growth to occur, there needs to be a base level of photosynthesis to meet the plant's metabolic needs. Once the needs of the plant are met, additional assimilates can be used for growth. To draw an analogy with people, we all have a base amount of calories we need every day to keep our heart pumping, keep our brain working, and maintain other essential functions. If we only get enough calories to cover these basic functions, we don't have any energy for other activities. Likewise, if any factor limits photosynthesis to the point where it doesn't cover the basic metabolic needs of the plant, growth will not occur.

The point at which a plant has met its metabolic needs and can move on to growth is called the compensation point. If photosynthesis is limited to the point where it is less than or equal to the compensation point, there will be no growth. Many environmental factors influence the rates of respiration and photosynthesis. For example, low light levels or a lack of water will prevent photosynthesis from reaching its potential. The rate of respiration is dependent on temperature: The higher the temperature, the faster the rate of respiration. In vining crops like tomatoes, when the temperature rises above 96°F (36°C), the rate of respiration surpasses the ability of the plant to photosynthesize. Time spent above this temperature will not result in growth because the plant is using all its energy for respiration.

This is why it's important to use the greenhouse to moderate low and high temperatures, and why we don't want nighttime temperatures to be hotter than they need to be. If night temperatures are excessive, the plant will use more energy than necessary, which will detract from the amount of energy available for plant growth.

Transpiration

Transpiration is the process of water moving through the plant from its roots to its leaves, from which it evaporates. When water passes out of the plant through open stomata in the leaves, it creates a deficit in the leaves; the plant must then pull water up from the roots to replace what was lost. Roughly 90 percent of the water entering most plants leaves through transpiration, with the remaining 10 percent staying in the plant.

The rate of transpiration is governed by temperature and humidity. Higher temperature and lower humidity lead to a faster rate of transpiration. This is where it becomes apparent how closely all three basic plant functions are connected. For example, if the plant isn't getting enough water to keep up with the rate of transpiration, stomata will close so the plant doesn't wilt. In addition to letting water out of the plant, stomata are what allow carbon dioxide into the plant for photosynthesis. So when transpiration isn't occurring, neither is photosynthesis. This is why environmental conditions need to be managed for the purpose of keeping plants active (see "Active

Root Pressure and Watering

Root pressure is a secondary way by which water is cycled through plants. If transpiration is the *pull* from water evaporating from the leaves, root pressure is the *push* from water absorbed by the roots. Root pressure isn't as strong a force as transpiration.

Root pressure can be manipulated through the amount of water applied, and the electrical conductivity (EC) of the irrigation water. A high amount of moisture and a low EC in the root zone will promote higher root pressure. This is because water wants to go from a lower to a higher salt concentration. A higher EC will make it more difficult for water to flow into the roots. An extreme example of this is when fertilizer burn results from excessive amounts of fertilizer salts around the roots, making it difficult for roots to draw water from the soil, to the point of damage.

High root pressure is undesirable, because it can promote guttation, meaning drops of moisture exuding from the leaves. Guttation is bad because it makes the leaves wet. The best way to avoid guttation is by avoiding overwatering, and not watering at night.

Figure 5.1. The water droplets along the edges of the tomato leaves are an example of guttation.

A rule of thumb is: Don't start watering until two hours after sunrise, stop watering two hours before sunset, and avoid watering overnight altogether. It's desirable to apply water only when the plant is active and the roots are taking up water. Otherwise, water tends to sit in the root zone. Standing water promotes the development of soil-borne diseases and excessive root pressure, which can lead to guttation and splitting in tomatoes. Cucumbers are also particularly prone to guttation, but other crops will experience this if root pressure gets high enough.

You can sometimes spot guttation through the sugars in the fluid it produces. In a best-case scenario, where the fluid has dried, compounds are left behind that are visible as a whitish film. In a worst-case scenario, such as in a very humid environment (which might contribute to guttation in the first place), the sugars in the fluid may begin to mold, which could lead to secondary infections.

The exception to the "no night watering" rule is in some arid environments, where very dry air may sometimes necessitate night watering. In these cases, it should be kept to the minimum necessary to avoid water stress.

Plants" on page 97). Because when plants aren't active, they're burning energy without making energy.

Factors Affecting Plant Growth in Protected Culture

The basis of protected culture is learning the ideal conditions for your crop species, and creating those conditions to the best of your ability given the capabilities of your structure. The main factors affecting crop growth in protected culture are light, water, carbon dioxide, temperature, humidity, and availability of nutrients. Manipulate the growing environment to keep these conditions as close to the ideal as possible.

Light

Since all plants use light to produce energy through photosynthesis, the amount of light is one of the most basic factors affecting plant growth. When it comes to extending the season, the question is whether your structure receives enough light to grow in the darkest part of the year. Even if you have plenty of heat, you need to make sure you have enough light.

The ability to measure the amount of light is important when deciding both where to situate a new greenhouse or hoophouse and whether or not to grow a particular crop at a certain time of year. When talking about light in regard to plants, we measure photosynthetically available radiation (PAR), because plants can't use all of the light humans can see in the visible spectrum. PAR describes the range of light that plants can use to photosynthesize (from 400 to 700 nanometers).

To assess the amount of light available in a given area, you need to know the daily light integral (DLI). DLI is like a rain gauge for light; it measures how much PAR an area receives. Just as you can look at a rain map and see how much water an area tends to receive, you can look at a DLI map and see how much PAR an area gets, averaged over a period of time. This is important because DLI takes into account cloud cover and other factors that reduce the amount of light that actually reaches the ground. For example, if you're considering building a greenhouse in an area that receives a lot of light but tends to be cloudy, a DLI map would warn you that a lot of the PAR would be blocked.

Generally speaking, the closer you are to the equator, the more solar radiation you'll receive, simply because more light hits that part of the globe than closer to the poles. This is why light levels have been more of a barrier to extended-season production in far-northern and far-southern areas. But a DLI map can show you more specifically which times of the year an area will or won't support plant growth. There may be times when your area receives borderline amounts of light for certain crops. If light levels are marginal, either you can grow a crop with lower light requirements – like lettuce or greens – or there may be enough light to support a seedling crop at the beginning or a topped crop at the end of the season.

The numbers on a DLI map express how much light you get on an average day in a given area. In protected culture, light is usually measured in moles of light falling on a given area per day, as in 15 moles (mol) per square meter (m^{-2}) per day (d^{-1}). In practice you will see this written as: 15 mol × m^{-2} × d^{-1}. DLI measures the cumulative amount of light an area receives; it's possible for a short, bright day to receive as much light as a long, overcast one. Most of the leafy crops can be produced with 12 to 14 mol × m^{-2} × d^{-1}, though they will grow faster with more light. The vining/fruiting crops need between 20 and 30 mol × m^{-2} × d^{-1} to produce well.

Look at a DLI map to find out what the wintertime average is for your area. You will see why growers planning on year-round greenhouse production of fruiting crops locate in the Southwest, or else install grow lights. Likewise, it's more viable to grow long-season fruiting crops with high light requirements in the Northeast than at the same latitude in the Pacific Northwest. The typically cloudy winter weather in certain areas of the Northwest precludes very long-season fruiting-crop production, even though you will have to use more heat in the Northeast. Lights for vegetable production (as opposed to seedling production) are such a big investment to install and run that for most growers, the best

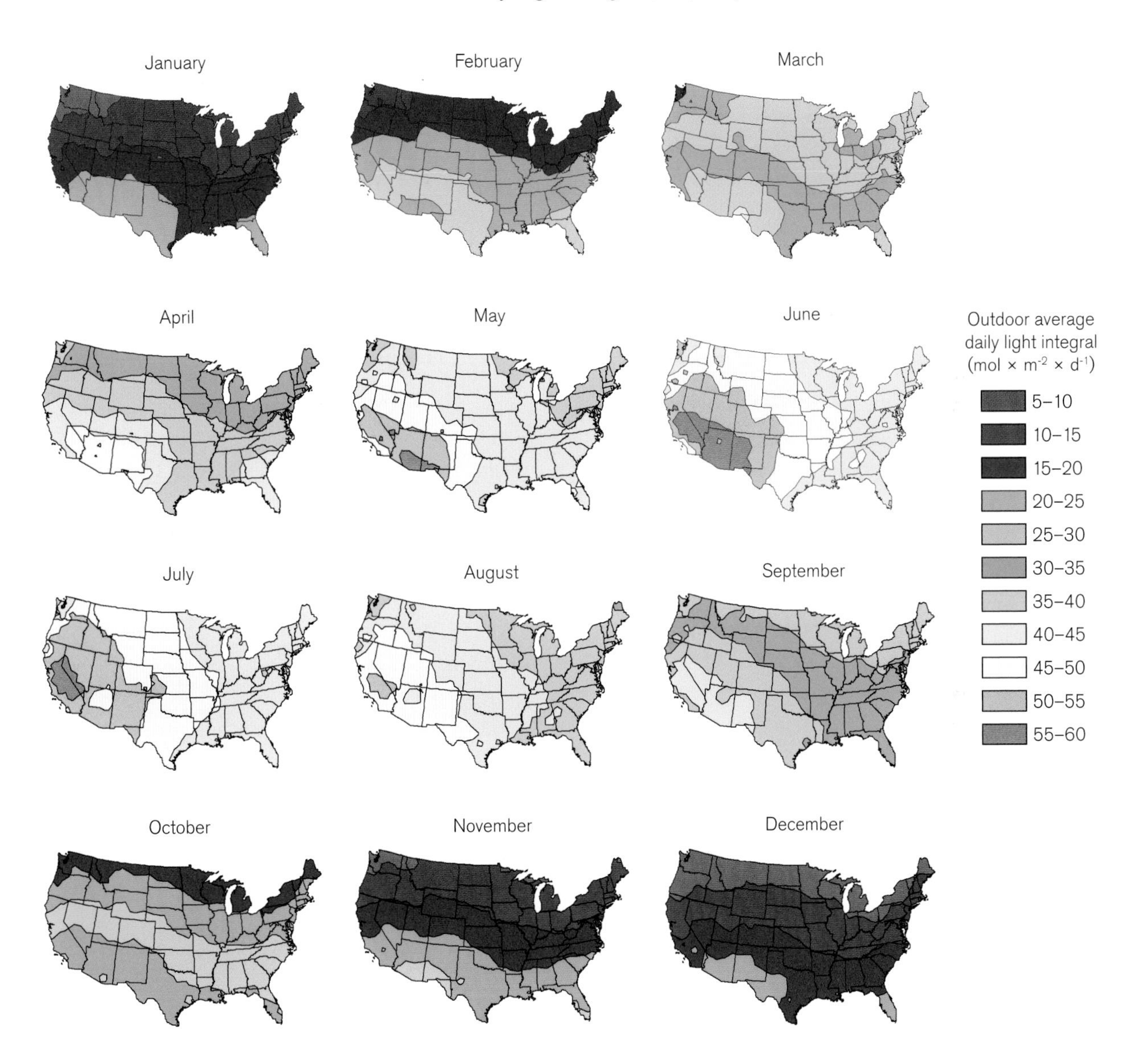

To estimate the DLI inside your greenhouse for a particular month:

1. Use a light sensor to determine light intensity outdoors at noon on a clear day.
2. Go into your greenhouse and take light intensity measurements at plant level.
3. Use these values to determine the percentage of light outdoors that reaches your crops. For example, if you measure 6,300 footcandles outside the greenhouse and an average value of 4,100 footcandles inside, your light transmission value is about 65 percent.
4. Multiply the DLI value indicated in the maps above by the transmission value to obtain the average DLI inside your greenhouse. For example, if your transmission value is 65 percent and the DLI for your location is 20 mol × m^{-2} × d^{-1}, then your average DLI that month is 13 mol × m^{-2} × d^{-1}.

Figure 5.2. A DLI map shows how much sunlight an area receives at different times of the year. Reproduced with permission from Jim Faust et al., "Mapping the Monthly Distribution of the Daily Light Integral Across the Contiguous United States," *HortTechnology* 12 (2002).

idea is to stick to growing crops that can be produced in their area using natural light.

Three principles allow growers to extend the season quite a bit without having to install lights in the main production area. The first is to propagate seedlings in the darkest part of the year with supplemental lighting. When they are small, the seedlings are grown at high density, so a much smaller area can be lit for propagation than for the same number of plants in production.

Second, seedlings need much less light to grow than do mature plants. Small seedlings may only require 6 to 9 mol $\times$ m^{-2} $\times$ d^{-1}, with light needs increasing as the plants get bigger. This is why it's possible to transplant seedlings into an unlighted production greenhouse, even in the north in January, and let the plants grow as the light levels increase. The low light levels at the beginning and end of the year may support seedling and leafy-crop growth though they won't support mature fruiting-crop production. One way to think of it is that often low light levels are enough to grow leaves but probably not fruit.

The last trick to get through dark times of the year without lighting is to increase spacing to reduce shading. In tomato production, it's fairly common to start the year with a certain number of stems, and develop an extra stem every few plants in the late spring. This increases the planting density to take advantage of high light levels in the summer. You can top this extra stem in the fall to stop its growth and reduce stem density again as light levels go back down. In this way you can manipulate plant density throughout the year to match light levels, with higher planting density corresponding to higher light levels.

Water

Make sure the plants' water needs are always met. When plants are drought-stressed, they close stomata to stop respiration; subsequently, growth also stops. Quality of water is important as well. Many soil growers don't check their irrigation source, but it's a good idea to do so. Some areas of the country may have water sources that are outside the ideal for plant growth. Water that is too acidic or basic may interfere with nutrient uptake, and will change the pH of the growing medium over time. If water is extremely basic or acidic, adjusting the pH isn't that expensive, and it's much cheaper than losing plant growth. There are in-line systems that can adjust water to a desired pH if necessary.

Carbon Dioxide

Low CO_2 levels can become a limiting factor for growth, especially during times of the year when the structure is kept closed to retain heat. Since plants consume carbon dioxide in the air during photosynthesis, levels can become depleted to the point that they slow growth when there isn't much air exchange. CO_2 should be thought of as fertilizer in the air, since it's consumed and depleted in the same way as nutrients in the soil.

Many growers supplement CO_2 because it can easily fall below the level needed for optimal growth. It's worth monitoring CO_2 levels in the greenhouse to know if this is happening. CO_2 management can take the form of venting specifically to introduce fresh air when levels are low, or supplementing CO_2 by adding it to the greenhouse.

One of the ways greenhouse growers boost yields is by augmenting CO_2 levels above the ambient level. For most greenhouse crops, boosting CO_2 to between 800 and 1,000 ppm increases photosynthesis by as much as 50 percent over ambient levels. Since CO_2 is only consumed during photosynthesis, augmentation only needs to occur during daylight hours.

Soil Fertility

Fertility is an important factor affecting plant growth because if a plant doesn't have enough nutrients, it cannot fulfill its growth potential. Soil growing is complicated because it can be difficult to determine how much fertility is available in the soil at any given time. This is one of the big reasons why large-scale commercial production of greenhouse vegetables has overwhelmingly moved to hydroponic production. But soil does have advantages, especially for smaller growers, including being a more forgiving medium, since it permits a larger root system with more soil volume per plant. This larger root area is capable of more buffering, or evening out and

Figure 5.3. We take advantage of greens' tolerance of much lower temperatures to plant a crop in the fall where we grow our fruiting crops in the summer. A layer of row cover inside the hoophouse provides a second layer of protection in the depths of winter.

Figure 5.4. Here are the results the following spring.

compensating for fluctuations in the amount of nutrients and water levels when abrupt changes occur. Plus, many organic certification programs, including those in Europe and some in North America, do not currently allow organic certification for hydroponic growing.

The starting point for getting good protected culture soil fertility is a good soil test. The specialized soil test used for greenhouse crops is the Saturated Media Extract (SME) test, sometimes called the Greenhouse Media test. Not every soil testing service offers this, but many university and commercial labs do. It's both inexpensive and well worth the money to get an idea of what's in your soil. Without a soil test, you'll have nothing to guide you in figuring how many nutrients are available in relation to what your plants need. Any applications of fertility without one are just guesses. An SME greenhouse soil test takes into account the larger amounts of nutrients that are necessary to feed a crop over an extended season.

The soil test results will provide recommendations based on the specific crop you intend to grow. But while they'll clue you in to what the crop needs, you'll still have to figure out what fertilizers to use, and in how many applications. Fertility needs for quick crops – most of the greens crops, for example – are easier to figure out; since their crop cycles are so much shorter, it may be possible to develop a standard pre-plant fertilizer that can be applied before every crop to feed it from transplant to maturity.

The challenge with long-season greenhouse crops is that it isn't possible to apply all the fertilizer necessary up front, at the beginning of the crop. Applying a whole season's worth of fertilizer would likely make the vining/fruiting crops excessively vegetative (see chapter 7) due to too much nitrogen in particular. The way to avoid this is to incorporate only a portion of the fertilizer into the soil before planting, and apply the remainder gradually throughout the season. Some growers apply half or a third up front and the rest over the course of the season. The two main ways of applying fertility after planting are side dressing solid nutrients to the root zone, and applying liquid nutrients through irrigation water, otherwise known as fertigation.

The advantage of using fertigation to meet supplemental fertility needs is that it's so easy. As long as you

Foliar Sprays

I am not a big fan of foliar feeding plants after the seedling stage, because this gives up two of the biggest protected culture advantages: dry leaves and clean produce. Wet leaves will increase the chances of developing foliar diseases, and foliar sprays have the potential to leave residue that will make produce look dirty.

have a fertilizer injector, it can be seamlessly incorporated into daily watering. Some growers make many small, dilute applications of fertilizer. Others will apply larger doses of fertilizer more infrequently. The downside to fertigation is that you're limited to fertilizers that flow through irrigation lines; anything that isn't completely soluble is off limits.

On the plus side for side dressing, you can use almost any fertilizer you want. However, side dressing requires more labor, and it can be tricky to activate solid fertilizers in the root zone of the plant. If fertilizer sits on top of the soil in a humid greenhouse environment, it may mold. On the other hand, if the greenhouse environment is very dry it may sit on top of the soil and stay too dry for the roots to penetrate it.

Finding the right frequency is a matter of matching the amount and kind of fertilizer to the needs of the crop. When applying fertilizer to soil crops, there are a few rules of thumb. If you're fertigating, get assurance from the manufacturer that the product will not clog drip tape or irrigation lines. Even seemingly soluble products can sometimes cause blockages. And clogged emitters on a hot day can be a serious problem, in addition to being a big waste of time.

If you're applying solid fertilizers, avoid pure applications of large amounts of blood meal that isn't thoroughly mixed in with other fertilizers. Its high nitrogen content can volatilize and cause damage to plants in confined environments. If you're using compost, only apply up to the point of meeting the crop's phosphorous needs. Using compost can lead to excess levels of phosphorous over time, so use other fertilizers to meet the rest of the needs of the crop. Since managing soil fertility can be tricky, do your own real-time monitoring of crop growth with crop registration (see chapter 7).

Tissue testing is another important way to see how well your crops' needs are being met. Just as soil testing will show you what's in the soil, tissue testing will tell you what's actually getting through to the plant. You can send a sample of leaves from a few different plants to a laboratory, which can tell you how the leaves' nutrient content compares with the ideal for the crop. Some growers only use tissue testing when there are physical symptoms that can't be linked to a particular pathogen, while others send off a tissue test every week or two to see what's going on in their plants. But regular tissue testing is well worth it in a greenhouse of any size, because it can tell you when plants are running low on a particular nutrient before deficiencies show up. This is important because by the time deficiencies show up, they will take a while to correct, and you have likely already lost some production. With regular tissue testing you can be proactive and head off major production problems before they occur.

Temperature

Temperature is one of the most basic influences on how any plant grows. Making the temperature more favorable than the prevailing outdoor temperature is one of the main reasons for wanting to grow in protected culture in the first place. Understanding how to control it is important for maximizing the health of your plants.

There are two important ways of thinking about temperature for understanding its effects on plants. Minimum and maximum temperatures express the range a certain species should be kept within temperature-wise. Twenty-four-hour average temperature is important because the average temperature over the course of a day's growing cycle has a big effect on the way a plant develops.

Minimum and Maximum Temperatures

The recommended minimum and maximum temperatures vary by crop, and form the boundaries within

Figure 5.5. This greenhouse in Florida has a retractable roof for maximum ventilation, and reinforced end walls to withstand hurricane-force winds.

which a species will be most productive. These are the highest and lowest temperatures at which the plant will stay happy. Most crops can tolerate temperatures well outside the minimum or maximum before dying, but plant growth and productivity will suffer during any time spent outside the ideal. These temperatures are useful for developing set points for thermostats. Though unheated-high-tunnel growers do not have as much control over the climate, they will still want to understand min/max temperature. Even in the most basic hoophouses, outdoor temperature will determine crop timing, tell you when to ventilate, and perhaps explain why crops are not doing well when min/max temperatures have been exceeded.

MAXIMUM TEMPERATURES

It's inevitable in most greenhouses that temperatures will go above the ideal at some point during the growing season. In heated greenhouses, cold stress can be eliminated by setting the thermostat at or above the minimum temperature, but in many areas in the summer, it's impossible to keep greenhouses below the maximum temperature, even with maximum ventilation and cooling.

Lowering nighttime temperature can help alleviate the effects of extreme daytime temperature to some degree, as explained in "Twenty-Four-Hour Average Temperature" on page 66. However, there is no way around the fact that crop growth will suffer when conditions are too hot.

When the maximum temperature is exceeded even slightly, growth efficiency may decline as the plant allocates more resources to respiring and keeping cool. If the temperature continues to rise and the maximum temperature is exceeded by a lot, water stress may occur as it becomes difficult for the plant to bring in enough

Figure 5.6. When grown at dense greenhouse spacing, double rows of vining crops like these tomatoes form a dense hedge of plants.

water to match the rate of transpiration. If water stress becomes severe, wilting may occur and the plant may close its stomata in order to stop further transpiration and water loss. Growth grinds to a halt under these conditions, since if the plant isn't transpiring it isn't bringing in water and nutrients for photosynthesis. Depending on the crop and conditions, excessive heat can kill the blossoms before they can set fruit.

It's impossible to put a point exactly on when plant death occurs. When a plant dies depends on a complex set of factors, including time spent at certain temperatures, available water, humidity, overall plant health, and more. If water stress is acute, it's possible for plants to lose individual leaves or parts of them that became too wilted to recover without losing the whole plant. Suffice it to say that you want to try to keep the growing environment below the maximum temperature, because effects on plant growth can be unpredictable once it's exceeded.

MINIMUM TEMPERATURES

The minimum temperature given for each crop is well above what the plant can tolerate and survive; rather, it's the lowest temperature that a crop can experience without a loss of productivity. Below this temperature, plants start to shut down, and not only will they lose productivity, but it will cost them precious growing time to get going again once they do warm up. What temperatures will actually cause crop damage and death are unpredictable. Frost can occur and cause severe damage to tender crops without actually reaching 32°F (0°C), depending on other atmospheric conditions.

Some greenhouse growers try to save money on fuel by running lower-than-ideal night temperatures. This can be penny wise and pound foolish, as they will save some money on the heating bill but will lose more money from lower production. Cold stress can hurt photosynthesis for multiple days following a cold night. If you're heating a greenhouse, paying the cost to keep it above the minimum temperature pays you back in extra production. Elimination of cold stress is one of the many reasons greenhouse growing is so much more productive than field and hoophouse growing.

Twenty-Four-Hour Average Temperature

The average temperature over a twenty-four-hour period determines how fast a crop will grow, with faster respiration at a higher temperature driving a faster rate of growth. Rate of growth determines how quickly new leaves, flowers, and roots are produced.

Plants respond more to the average temperature than to extremes. That's why at the beginning of the climate section for each individual crop chapter (chapters 9–13), I provide a recommended average temperature for each crop. This is the average temperature that will maximize yields of fruit or leaves, depending on the crop. Ideally you want day and night temperatures to average out to this number.

Using the average temperature in conjunction with the min/max temperatures for each crop will give you

Using Minimum Temperatures for Hoophouse Crop Scheduling

In all but the mildest of areas, cold stress is a fact of life for hoophouse growers. Though they heat up a lot when the sun is out, single- or double-layer hoophouses don't stay much above the ambient air temperature at night. In fact, if the nighttime low temperature holds for an extended period of time, the temperature in the hoophouse may get as low as it is outside. Hoophouse growers can plant earlier than field growers, but not by a whole lot. At the very least, growers using unheated structures need to wait until danger of frost is past to plant sensitive plants like the fruiting crops in this book. It's not worth pushing it, trying to plant on the very first day of the season with a frost-free night.

From experience, I can say that by the time you get your hoophouse planted, meteorologists will have revised the forecast back into freezing territory. Also, though the structure may warm up a lot by day, it's still going to be very cold at night, so even if you plant just after the last frost date, the plants are going to experience a lot of cold stress and not grow very fast anyway. If you have the plants in a heated propagation area, it's a good idea to leave them there where they will continue to grow actively rather than moving them into a hoophouse where they will stall out until the weather improves. It's your job to keep the plants in continual growth, because once they have stalled out for any reason it will take them a while to recover.

Hoophouse growers should use nighttime lows as a basis for crop scheduling, and wait until temperatures are well out of the damaging range to plant sensitive crops.

an idea how much latitude you have temperature-wise to even out extreme temperatures. The most common example of this is using cooler night temperatures to balance out hotter-than-desirable day temperatures.

For example, let's say you want to run an average temperature of 70°F (21°C) in a cucumber crop. You know your maximum temperature is 86°F (30°C) and your minimum, 65°F (18°C). You can make up for a hotter-than-desirable day by making night temperatures a little cooler than usual. But you can't make up for a 100°F (38°C) day by making it 40°F (4°C) overnight. The plants would shut down if it got that cold. Based on your minimum temperature, the coolest you would ever want to go on a cuke crop would be 65°F (18°C). So minimum and maximum temperatures set the parameters within which to try to create a given average temperature.

In addition to determining how fast the plant grows, the temperature under which plant growth occurs plays a part in determining the *nature* of the growth for the fruiting crops (tomatoes, cucumbers, peppers, and eggplant). At lower twenty-four-hour average temperatures, internodes are closer together, flowers are larger, and stems and leaves are thicker. Growth initiated at higher twenty-four-hour average temperatures tends to have the opposite characteristics: Internodes are farther apart, and flowers, stems, and leaves are smaller. Other factors figure into the plant's morphology; how generative or vegetative a plant is will affect these same plant features (see "Crop Steering for Generative or Vegetative Growth," chapter 7).

I've heard a lot of people say that the twenty-four-hour average temperature seems lower than they would have expected for each crop. Keep in mind that it's an average, and that daytime temperatures will be higher and nighttime temperatures lower. Temperatures should be adjusted on a daily basis in reaction to weather conditions. For example, on a day when daytime temperatures get too high, nighttime temperatures can be lowered (though not below the minimum temperature for the crop) to even out the effect on the twenty-four-hour

Day/Night Temperature Differential

It's not enough to simply set the temperature controls to the ideal average temperature. Especially for the fruiting crops, it's desirable to have a day/night temperature differential. Even if you could somehow maintain a flat temperature profile with no day/night variation, this would have an extremely vegetative effect on fruiting crops. Because tomato crops have been so extensively studied, they offer the best example of why this is the case.

All other things being equal, the fastest way to grow a tomato plant would be to set day and night temperatures to 77°F (25°C). But this would result in an excess of leaf and shoot growth at the expense of fruit production (see "Crop Steering for Generative or Vegetative Growth," chapter 7). So it *would* be the fastest way to grow the plant, but not the fruit. The day/night temperature differential in each crop's temperature recommendations has been developed to maximize the speed of crop growth without compromising fruit production.

average. This is where good temperature sensors and computer control are handy, because the computer can calculate daily average temperatures in real time for you to base your heating and cooling decisions upon.

Relative Humidity and Vapor Pressure Deficit

It's just as important to manage the water in the air as it is to manage the water in the soil. Extremely high or low humidity is bad for plant growth. Relative humidity (RH) and vapor pressure deficit (VPD) are two different ways of expressing the same thing: how much moisture is in the air. But whereas RH measures how full of moisture the air is, VPD measures how much capacity the air has to take on additional moisture.

Understanding RH and VPD is important for protected growers because along with increasing control over the climate comes control over humidity. The amount of moisture in the air has a big effect on plant growth, so manipulating humidity is an important part of optimizing plant growth. The ability to manipulate humidity to keep it within an optimal range is a huge benefit of having a heated greenhouse. This is particularly useful early and late in the season when structures are kept closed to preserve heat, and humidity can build up to very high levels.

Relative humidity is a percentage that expresses how much moisture is in the air versus how much moisture the air can hold. For example, an RH of 50 percent means that the air is halfway saturated with water vapor and can hold twice as much as it is currently holding. The higher the number, the closer the air is to being at capacity. RH is somewhat imprecise for determining the impact of a given amount of moisture in the air on plant growth. This is because relative humidity is highly dependent on air temperature. The water-holding capacity of air roughly doubles with each 20°F (10°C) increase in temperature. The RH of a given body of air increases if the temperature is lowered, and decreases if the temperature is raised. Picture cold air as a small bucket, with little capacity for moisture, and hot air as a large bucket. The hotter the air, the bigger the bucket gets. In fact, air at 77°F (25°C) can hold more than twice as much water as air at 50°F (10°C).

Vapor pressure deficit is essentially a temperature-compensated way of expressing the moisture carrying capacity of air. Instead of expressing how much water is in the air, VPD is an expression of how much capacity a body of air has to absorb more moisture. VPD and RH have an inverse relationship. When RH is high, VPD is low.

VPD can be expressed in kiloPascals (kPa) or millibars (mbar), though mbar is more common in greenhouse uses. Either format is an expression of the pressure that water vapor in the air exerts. VPD is more accurate than RH in measuring the effect of humidity in the air at

Figure 5.7. As plants grow up they shade each other, which is how to maximize the number of plants in a structure. The plants don't suffer, since most of the photosynthesis occurs in the upper leaves, and on a bright sunny day they can get more light than they can use.

any given temperature, because it's an expression of the transpirational pull, or the strength of the movement of water from the leaf into the atmosphere.

Since the interior of a living leaf is always wet, VPD reflects the difference between the vapor pressure in the leaf and in the surrounding atmosphere. This is why little transpiration occurs when VPD is very low. In other words, low VPD means there isn't much difference between the amount of water inside and outside the leaf. If it's very humid, there isn't much draw for the moisture

Figure 5.8. The piece of equipment in this teaching greenhouse at the University of Arizona is a carbon dioxide generator. When the amount of CO_2 goes below a set level, it turns on, giving off CO_2 and heat.

inside the leaf to come out. VPD can be calculated using a chart showing the value at each temperature and relative humidity. Many climate control programs will calculate VPD for you.

The bottom line for plants is that as RH gets higher and VPD gets lower, transpiration will become increasingly difficult. Conversely, when RH gets low and VPD gets high, this drives a higher transpiration rate. Plant transpiration is active in that a plant can open or close stomata to start or stop it, but it's passive in the sense that it depends on the air being able to accept moisture from the plant. It can't just spit moisture out the stomata if the air becomes too saturated to absorb the moisture the plant is trying to release. In the most extreme scenario, if RH is 100 percent and the air is saturated, transpiration won't be possible until the RH goes down and VPD goes up. What we're looking for in protected culture is a happy medium where transpiration is active but not too fast. For most greenhouse crops, this happy medium is a VPD between 4 and 8 mbar, or an RH between 70 and 80 percent.

As for the relationship between transpiration and RH/VPD, anything that causes the process of transpiration to shut down must be avoided, because when plants are not transpiring, they aren't cycling water and nutrients, and they aren't growing. Plants stop transpiring when their stomata have closed to conserve moisture. Even if their stomata are open, plants may not be able to transpire if RH is very high and VPD very low.

Even when plants can respire, RH or VPD outside the ideal range may cause respiration that's too fast or too slow. High humidity slows transpiration, and low humidity speeds it up. The closer the air is to being saturated, the more difficult it becomes for plants to release water into the air. Since water is one of the by-products of photosynthesis, not being able to release it slows photosynthesis. Sluggish transpiration will also cause slowed growth, because plants aren't cycling water and nutrients through the plant fast enough.

At the opposite end of the spectrum, if the air is so dry that roots can't keep up with the rate moisture is leaving the leaves, stomata may close to try to keep the plant from losing too much moisture and wilting. Since transpiration cools the plant, closed stomata during hot parts of the day can result in overheating and damage to the plant.

The effects of stopped transpiration are so dramatic that they're apparent to the touch. Under most circumstances, if you touch the leaf of a plant in a greenhouse or hoophouse it will feel cooler than the surrounding air, because the evaporation from transpiration is cooling the leaf. If plants feel hot, it's likely that transpiration has stopped. Regardless of what's causing it, a hot plant is a bad sign. It could mean that humidity is so high that it's preventing transpiration from occurring even though stomata are open. Or it could mean that stomata have closed as a reaction to water stress, and the plants are on the verge of wilting.

Prevailing VPD conditions have a strong effect over plant morphology. For example, plants grown in more humid northern North America will look different from the same varieties grown in the dry desert Southwest. Plants grown under humid conditions will have larger, thicker leaves and stems. Plants grown under dry air conditions will have smaller, thinner leaves and stems.

When RH is chronically too high and VPD too low, plants can have weak root systems, as the plant is not stimulated to produce a lot of roots to feed transpiration. A temporary condition called lazy root can occur when transpiration is slowed for a prolonged period and then picks back up again; for example, during an overcast period followed by hot and sunny weather. This condition can be the culprit behind temporary wilting following a cloudy spell, while the roots catch up with the plant's demand for water. The good news is that plants usually outgrow lazy root as long as other growing conditions are normal.

The increased control of protected culture is only useful if you understand how to manipulate the environment and the plants grown in it. Knowing how to help the plants grow as well as possible requires understanding the many factors that impact growth. The actions we take in protected culture are based on the many fascinating ways that plants are influenced by their environments. This summary of plant biology as it relates to protected culture has laid the foundation for many of the concepts yet to come in the book.

PART TWO

Crop-Specific Best Practices

CHAPTER SIX

Propagation, Pruning, and Trellising

The techniques used for growing tomatoes, cucumbers, peppers, and eggplant in protected culture are very similar. One of the nice things about these vining/fruiting crops is that once you learn how to grow one, you can apply what you know – in modified form – to the others. Since all four crops can be grown with the same overhead trellis wire spacing (all can be grown in 2-foot/60-cm-wide double rows), it makes them suitable for mixing and matching different vining crops from year to year depending on demand or crop rotation. The next three chapters are the basics you can apply to all of them; the specifics of each crop can be found after that, in their respective chapters.

Seeds

Since all the plants covered in this book are started from seed, it's important to understand that good plants come from good seed; buy your seed from a reputable source. There are a lot of diseases that can be transmitted by seed, and most of the companies selling seeds specifically for greenhouse production do a good job of sourcing and testing seeds to make sure they're free of pathogens.

If you save any of your own seed, or source from anywhere you don't completely trust, find out what diseases may be seedborne for the crop species you are growing. There are a number of hot-water or chemical treatment methods that can eliminate or reduce seedborne pathogens.

Good Seed and Plant Practices (GSPP) is an international accreditation program designed to combat one of the most problematic seedborne diseases of tomatoes: bacterial canker (*Clavibacter michiganensis* subsp. *michiganensis*). The program was created in response to very serious losses suffered in commercial greenhouses due to infected seeds. When seed is GSPP-certified, it means it was grown under conditions designed to minimize pathogens. Certified seed is produced in carefully controlled greenhouses to isolate the seed crop from any risk factors that may exist in the environment. Then to make sure, seed from each lot is tested for the presence of bacterial canker.

Though this system was originally devised to protect tomato seed, other fruiting crops not susceptible to bacterial canker are being produced with GSPP certification, to give growers confidence that the seed is free from pathogens. To date, GSPP certification is usually only sought for seeds of varieties specifically for greenhouse production.

Why Greenhouse Seed Is So Expensive

I'm often asked why the seed of greenhouse vegetable varieties is so expensive. The answer has several parts but one of them is GSPP. It is more expensive to produce seeds in a tightly controlled greenhouse, and then have them pathogen-tested, than to produce them in the open field.

But GSPP certification, and buying quality seed in general, is cheap insurance compared with the expense of losses due to disease, especially when you consider that many diseases can spread quickly from plant to plant. Even just a few sick plants grown from infected seeds can be the start of a serious pathogen problem or even crop failure within a greenhouse.

Other parts of the greenhouse seed price equation include the fact that all the breeding, testing, and development of greenhouse varieties takes place in greenhouses, which is more expensive than field space. Just as field varieties need to be developed in the field, greenhouse varieties need to be tested where they're intended to grow if results are to be reliable. In addition, the greenhouse seed market is a smaller, much more specialized market than that for field seed. The demand for greenhouse seed worldwide is a fraction of the demand for field seed, so each greenhouse seed bears a higher percentage of the burden to pay the breeder back for development costs.

The higher cost of greenhouse seed is balanced by the fact that you should be getting a higher return per seed from greenhouse seed. You can get a much higher yield over a longer season with a greenhouse variety than you can with a field variety. My hoophouse tomato plants are a good illustration of this. By grafting, I use two seeds to make one plant: the rootstock and the top variety. From those two seeds, I will make a double-leader plant. In a good year, even in an unheated hoophouse in Maine, I can get 30 pounds (14 kg) of fruit off each leader. At a low average price of $2.50 per pound, that one tomato plant has the potential to gross at least $150 for me. If you get just one more piece of fruit off a greenhouse variety than you would from a field variety, that should cover the increased cost; anything beyond that is profit.

You can find out if paying the price for greenhouse seed is worth it by growing a row of a greenhouse variety and a row of a field variety, keeping track of yield and prices, and as a winter project compare the seed-price-to-revenue ratio of each. If the field variety comes out ahead, maybe it's not worth it for you to pay the premium for a greenhouse variety. But for most farms, seed cost is far down on the list of expenses. If you find that there are advantages to growing a greenhouse variety, it's probably penny wise and pound foolish to try to save money by using field varieties. Whether they save labor, resist a disease, or simply provide a higher yield, using the best varieties for your growing environment will pay off in the long run.

Choosing Varieties: Protected Culture Traits

There are a number of specialized adaptations that help varieties thrive in protected culture. Plant breeders select for traits that are advantageous for growing in a hoophouse or greenhouse. Keep these traits in mind when choosing which varieties to grow.

So what do you get for all the fancy breeding and development work on protected culture varieties? As mentioned above, greenhouse vegetable varieties come out of completely different breeding programs than their field counterparts. Protected breeding programs have been separate from field programs for decades, dating back to the beginnings of commercial greenhouse production, as growers identified different needs for plants in protected environments. One of the most basic requirements for a variety is for it to thrive under protected conditions. There are a lot of differences from field growing simply because you are under cover. Light levels are not quite as high, and are more diffused than in the field, since all coverings block some amount of light.

Another important aspect of breeding for greenhouse varieties is plant habit, sometimes called plant architecture. This varies based on the crop being grown, but greenhouse crops frequently have ideal plant types that differ from the field version of each crop. For example, breeding may focus on making the plant more compact, while keeping the plant habit open enough for good airflow, so they can be planted densely.

Resistance to common greenhouse pathogens is another basic breeding goal. There is some overlap, but many of the important disease resistances are different between field and protected crops. There are a number of diseases seldom seen in the field that are common in the greenhouse.

With all the other breeding goals, flavor sometimes takes a backseat. It's always a good idea to try a new variety in a smaller quantity before going into full production, just in case it's not as good as you think it's going to be. This includes all aspects of growing, but especially flavor. Flavor is so subjective that it's hard to take someone else's word for it. So whatever your flavor standards are, make sure a new variety meets them before growing a lot of it. It's tempting to put a list of my favorite varieties here, but they change so quickly – and opinions are so subjective – that I'm going to resist the temptation.

Propagation

Tomatoes, cucumbers, peppers, and eggplant are propagated in a similar manner. Propagation is an extremely important part of raising crops, because if your plants don't get off to a good start, problems that arise in the beginning may persist through the entire life of the crop.

Propagation Mediums and Cell Size

Before a single seed is planted, two big decisions need to be made: what size cell to use, and what medium to put the seed in.

As for mediums, biodegradability is a constraint for soil growers, who want something that will become incorporated into the soil when it breaks down. Many soil growers use homemade soil mixes or commercially available potting mixes. Soil growers can also use cocoa coir, peat-based foam, or other biodegradable mediums for starting plants to be transplanted into the soil.

Soil-based mediums may carry varying amounts of fertility, pathogens, and weed seeds. I've noticed this problem when my usual potting mix turned out to be stronger or weaker than usual. For this reason, many soil growers have their propagation mixes tested before use, even if they come

Figure 6.1. In this picture of our nursery, mesh is keeping most insects out of the structure, yellow sticky cards help us monitor what's in the structure, and HAF fans maintain good airflow around tightly spaced seedlings.

Figure 6.2. Fiber pots allow you to grow a larger plant than cell flats would permit. They also make it easy to transplant a larger plant, since you can plant them pot and all, without having to remove them before putting them in the ground.

from a commercial supplier. Seedlings are more sensitive to pathogens and fertility imbalances than larger plants, so it's important to make sure they start off right. When you think about the cost of the seed plus the delay caused by a failed planting, successful propagation becomes very important. An alternative to testing batches of potting soil is to do a test-grow prior to starting the first round of seedlings, to make sure the potting soil performs as expected and that the seedlings aren't affected by pathogens.

Though soil growers have the flexibility to use any cell size that they can find a tray to hold or a soil blocker to make, cell size is usually a function of whether space or labor is at more of a premium (I use the term *cell size* to refer to plants started in soil, since the empty cell determines the size of the root ball; the term *block* is used to refer to the larger containers or soil blocks a cell may be transplanted into for the seedling to grow larger). If you're short on propagation space, you can use the smallest size possible to germinate seeds, and then transplant to a larger size as needed. If you have plenty of space and are short on labor, a more time-efficient solution is to germinate seeds in a larger cell, to reduce or eliminate the number of times you have to repot to a larger size.

But in general, seeds are usually germinated in cells or plugs much smaller than will be transplanted for production, because greenhouse growers tend to use a much larger transplant size than field growers. If seeds were started in the block they were to be transplanted in, it would waste space and cost more to heat during propagation; in addition, seeds frequently germinate better in smaller cells. It's harder to manage the watering needs of a very small plant in a large cell. The exception to this is the leafy crops, whose growers have the option of seeding straight into the size used in production.

Propagation Space

Greenhouse growers use large transplant sizes because the longer seedlings stay in propagation at high density, the less time the entire greenhouse has to be heated. For this reason, some growers have separate propagation houses that are smaller than their production houses, just for producing seedlings. Another option is to put up a plastic curtain and divide off a section of the production house to use for propagation so you don't have to heat the whole area. Many growers opt to pay a propagator to grow the seedlings for them, since propagation is such a specialized process, and quality transplants are so important to having good crops. This is especially true for long-season crops, like heated-greenhouse peppers, that need to have seedlings growing before the end of the previous crop.

For growers not using heat, instead of trying to plant early into a hoophouse where temperatures may be too cold, let the transplant get big in a heated nursery. More mature transplants will begin to yield more quickly and reduce the amount of time that the structure is not producing. Larger transplants are possible for greenhouse growers because unlike field growers, they don't have to contend with wind and rain beating up large transplants.

Fruiting-crop growers do not have this option. Seeds do not germinate and grow well in the large blocks that are used to produce fruiting-crop transplants, plus it would be wasteful of space to devote 16 in^2 (100 cm^2) or more to each germinating seed. The ultimate in compact planting for solanaceous crops is a twenty-row flat, which has the same dimensions as a standard 10 × 20-plug flat but instead of cells there are twenty rows. Seeds can be planted densely (up to twenty per row or so) and plucked out and transplanted into cells shortly after germination. This is extremely conservative of space though it does result in more labor transplanting the seedlings into cells. This does not work as well for cucumbers as for the solanaceous crops, however, since cucurbits grow faster in the seedling stage and resent being uprooted. It is recommended to start cucumbers in cells rather than a twenty-row flat.

To attain a large transplant size, soil growers may use fairly large pots with as much or more volume than a 4-inch (10 cm) cubic block, to transplant their seedlings into. I have seen soil growers using pots up to 0.5-gallon (1.9 l) size to produce extra-large transplants. A pot that holds 1 quart (1 l) of soil is a good size for growing a big transplant. Just make sure the root volume is proportional to the size transplant you want to grow. Big transplants are great, but not if they're root-bound.

Germination

From troubleshooting customer problems at a seed company, I know that a lot of the problems encountered with germination have to do with temperature. Pay attention to the ideal germination temperatures for each species and maintain them as consistently as possible for good germination.

One setup for poor germination I have seen many times: thinking that greenhouse temperatures are adequate for germination. Though daytime temperatures may be high enough, they have to stay constant for good germination. The nighttime temperature dip in most greenhouses is low enough to throw germination off. This is why building or purchasing a germinator is the ideal for germinating any crop. A germinator is basically a temperature-controlled box with shelves for flats of germinating seeds. It can be set right at the desired temperature for whatever you're growing. You can make one fairly cheaply by putting a heater in the bottom of a box and regulating the temperature with a thermostat.

There are also commercial models that have lights above each shelf, since the biggest pitfall of a germinator is that it's easy to leave the plants inside for too long after germination. Seedlings become stretched very quickly in the dark at high temperatures.

The second best option after a germinator is to use bottom heat from heat mats or hot water to germinate seeds. Bottom heat is more vulnerable to fluctuations in the surrounding environment than a germinator, but it's still a very good option.

Regardless of medium, plant one seed per cell, and cover with fine- or medium-grade vermiculite. Water the seeds in gently, taking care not to wash them out of their cells. After watering, cover with a plastic sheet to keep the seeds from drying out. If the flats will be going into a germination chamber, cover them with clear domes or plastic sheets. Covering with something clear will make it much quicker to check the germinator without having to take the covers off. If they're going onto heat mats on bench tops, cover with white domes or plastic sheets. Transparent domes in a greenhouse will get too hot and cook the seeds.

The highest, fastest, and most uniform germination is produced by maintaining the right temperature consistently from planting until emergence. Some varieties can be finicky germinators if they're not given this treatment.

If you use a germination chamber without lights, keep a close eye on germination, because sprouts will stretch and start getting leggy immediately if they are hot and don't have any light. This can be a big problem, because stems that stretch a lot in the beginning may lose the ability to expand to a normal size later on. As soon as you see the first sprouts emerge, move the flats to heat mats with plenty of light. That way the first to sprout will not get leggy, and those not yet sprouted will continue the germination process.

Seedlings

As mentioned above, many fruiting-crop growers like to produce a larger seedling than is typically used in the field. Growers who don't have a separate nursery may

Figure 6.3. These seedlings are ready to go into the hoophouse.

Figure 6.4. A freshly transplanted soil-grown grafted tomato plant. The rubber band was hooked onto the bamboo skewer, wrapped around the stem, and hooked on the other side to support the plant until it can be trellised. Note how the original plug was transplanted into the block, and the block transplanted into the soil a little higher than the original soil line to keep the graft union well above the soil line.

Transplanting Fruiting Crops

Hardening off seedlings isn't usually necessary in protected culture production the way it is for transplants going from nursery greenhouses to the field. Since the transplants will be moving from a propagation greenhouse to production greenhouse (or from a sectioned-off part of the greenhouse to the larger structure), the shock of going from a protected environment into the open field isn't an issue.

If soil growers are using blocks of cocoa coir or another solid medium that doesn't disintegrate over the course of the season, they can set the blocks on the soil surface and water them in with dripper stakes without burying the blocks to save time. This is more like the procedure for transplanting into hydroponic systems (see appendix A). Otherwise, they have to dig a hole large enough to accommodate the root ball. It's important to start irrigating fruiting crops shortly after transplanting to encourage rapid root establishment.

heat a partial bay of a larger greenhouse and use it as a nursery by putting a curtain up to separate it from the rest of the greenhouse. This minimizes the amount of time the entire greenhouse needs to be heated before crops become productive and start paying for themselves.

Often the longer nursery periods for vining crops mean they reach the stage where they flop over before being transplanted into the production area. Deal with this by sticking a bamboo or wooden stake into the soil next to the plant (bamboo shish kebab skewers work well for this), and loosely rubber-band the plant to the stake. If the leaves are too big for you to put a rubber band over the top of the plant, you can hook the rubber band onto the tip of the stake, wrap it around the stem – being careful not to encircle any leaves – and hook the other end of the band over the stake as well. Get rubber bands that are tight enough to hold the plant up but loose enough not to cut into the stem. These do not usually need to be removed, as the high light and temperature in the greenhouse will quickly degrade the rubber until it falls apart.

Crop Spacing

One basic greenhouse spacing layout can accommodate all the popular vining/fruiting crops. (I'll get into the spacing details for each individual crop in their respective chapters.) A big benefit of standardizing layout is that you can then grow any crop in any part of the greenhouse without having to change the trellising hardware. Installing trellising is time consuming and you don't want to have to redo it every year. It's ideal to pick one spacing that works for all your crops and stick with it.

The system that I see most often, and that I have used myself, is double rows of plants 2 feet (60 cm) apart with a 3-foot (90 cm) walkway between them. The hardware for this setup involves installing two wires 2 feet apart above where you want your double rows to be, then leaving 3 feet for a walkway until the next set of wires. This results in a very simple setup with two wires 2 feet apart, 3 feet for a walkway, and another two wires 2 feet apart, repeated across the whole greenhouse.

Some people feel that this spacing is tighter than they would like, and that they can achieve the same plant densities by leaving 4 feet (1.2 m) for a path and crowding the plants more in the row. But my best advice is to space the wires as tightly as you would ever want them, since you can always decrease density later by increasing in-row spacing.

Yield will be higher when plants are spaced more closely together simply from having more plants in the structure, which is why you want to space plants as tightly as possible while still keeping them healthy. The spacings recommended in this book are fairly standard, and are how Dutch growers get their plants as dense as possible without compromising plant health.

Calculating Plant Density

Calculating crop spacing is an important way to figure out the density of your greenhouse planting. Only

Intercropping

Intercropping is the practice of planting new seedlings in with older plants in the greenhouse. It's a strategy to eliminate the break in production caused at the end of the season between when the old crop is cleaned out and the new crop is put in. By the time the older plants are removed, the younger plants are in production, creating a constant year-round production cycle. This may be an advantage for keeping certain markets and providing year-round income.

While intercropping represents the ultimate expression of liberating growing space from the limitations of weather, there are serious disadvantages. Pest and disease problems from the old crop will carry over into the new crop. Also, the farther away you are from the equator, the more likely it is that you'll need supplemental lighting to interplant, which adds to the expense and complication of this technique. And it makes cleaning out the old plants more difficult, since the new plants are in the way.

This advanced technique has been problematic even for experienced growers.

looking at in-row spacing can be misleading, because even a tight in-row spacing can lead to a low planting density if the rows are far apart.

The easiest way to calculate planting density is to draw a map of your greenhouse or hoophouse, and figure out how many vines are in the growing area. A vine could be an individual plant in the case of single-stemmed plants, or each individual leader in the case of multistemmed plants. Don't include walkways larger than regular pathways between rows, utility areas, or any other space that has a lower-than-normal planting density. Though that does reduce the overall planting density of your structure, including these types of areas will throw off your calculations and make your planting seem less dense than it is. What you really want to know is how dense it is for a plant in the middle of a typical row.

For converting calculations in square feet to square meters, divide the number of square feet of growing area by 9. Then divide the total number of plants by the number of square meters to get the number of plants per square meter. It's not important that the number you get be exactly the plants/square meter target you are aiming for as long as it's pretty close. Differences in the way you do your layout and account for space will result in slight differences in the outcome, so this is more important as a ballpark figure to see roughly how dense your planting is.

Heads and Leaders

It's common for growers to refer to the top part of a vining/fruiting-crop plant as the head. This takes into account all the leaves, flower clusters, and shoots in the top 4 to 8 inches (10–20 cm) of the plant. You can tell a lot about the state of the plant by looking at the head, where the new growth is found. For example, a thin stem with a flower cluster close to the head indicates a plant in a generative state, whereas a thick stem with a lot of growth above the flower cluster indicates a more vegetative plant state (for more, see chapter 7). It's also common to refer to each vine as a leader, as in *single-* or *double-leader plant*.

See figure 6.5 for the most common spacings to achieve the most common densities.

You may be able to change planting density over the course of the season, as the amount of light available in a greenhouse changes. This is especially true in long-season

Recommended Spacings for Vining Crops

All the vining crops in this book can be grown using a system of double rows trellised to overhead wires spaced 2 feet (60 cm) apart, alternating with 3-foot-wide (90 cm) walkways. This standardization of layouts is advantageous when you're rotating crops through different places in the greenhouse over time without having to change trellising. Figure 6.5 shows how to space plants for the recommended planting densities. Adjust as needed. Due to their smaller leaves and more open plants, smaller-fruited varieties can often be crowded even more densely to increase their typically low yields. If you need more airflow or wish to make up for low light levels, you can space plants farther apart in the row. See each crop section for more detailed pruning and trellising information.

Tomatoes. In this drawing, the plants are spaced a foot apart, in the middle of the row, with two leaders going from each plant up to the trellis wire, and all the vines spaced evenly 1 foot (30 cm) apart along the wire. Plants can be lowered and leaned as they outgrow the height of the wire. Variations: Single-leader plants could be planted in double rows a foot apart to achieve the same spacing.

Cucumbers, umbrella-style. Plants are spaced 18 inches (45 cm) apart in a single row, with a single vine per plant going to alternate sides of the double row. When the vine reaches the top wire, it is topped so two shoots grow out to either side of the wire. When they get long enough, the side shoots are guided over the

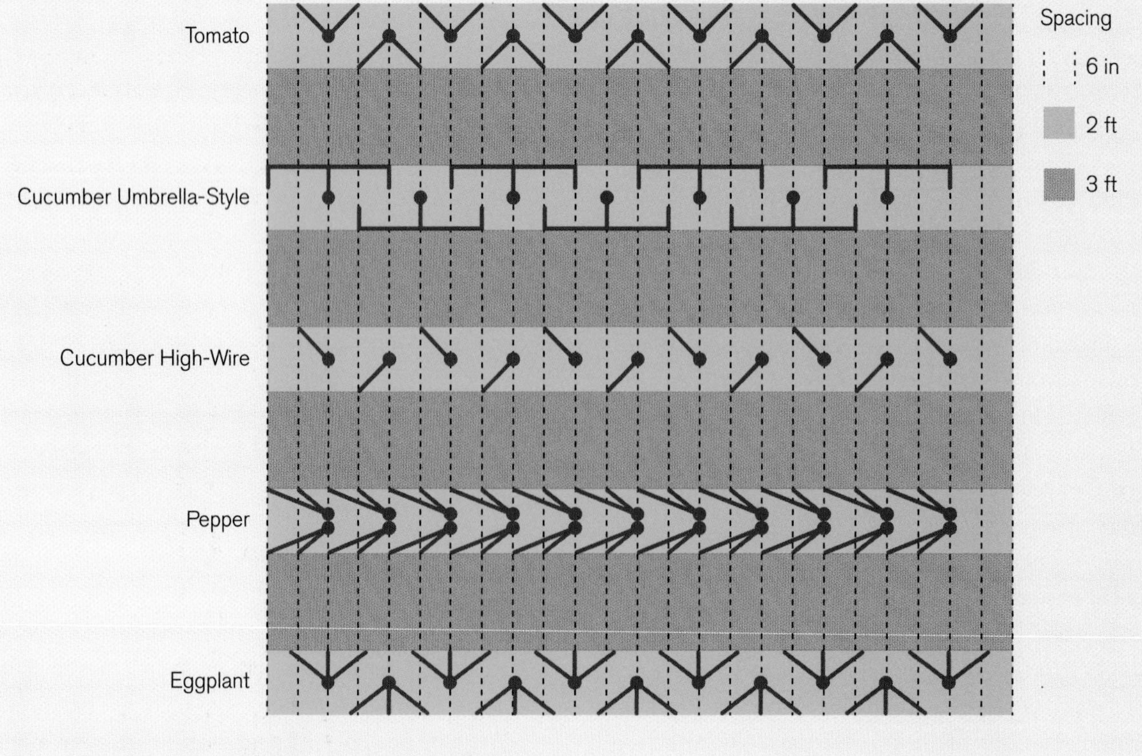

Figure 6.5. Recommended Spacings for Vining Crops

trellis wire for support and allowed to grow out roughly 18 inches to each side of the main stem, filling the space between vines.

Cucumbers, high-wire. Single-leader plants can be spaced a foot apart in the center of the row, with vines going to alternating sides, so vines are spaced out 2 feet (60 cm) apart along the wire. Plants are lowered and leaned as they outgrow the height of the wire. Variations: In summertime/high-light growing conditions, plants can be laid out as densely as double rows with a plant every foot (30 cm), with vines spaced a foot apart along the wire.

Peppers. Plants are put in double rows a foot (30 cm) apart, with heads spaced out evenly 6 inches (15 cm) apart along the wire. Variations: If you're using plants with four heads, you could put plants in double rows 2 feet (60 cm) apart, or single rows a foot apart, with heads spaced out evenly 6 inches apart along the wire.

Eggplant. Plants are put a foot apart in the center of the row; two leaders are initially developed and spaced out a foot apart along the wire. When light levels are good, a third leader is developed from each plant and spaced evenly along the wire. Variations: Up to four leaders can be developed at the same spacing to achieve a higher planting density. This may work well for lower greenhouses, as four leaders will grow up to the wire more slowly.

greenhouse production where the season begins and ends in dark times of the year with short days. Hoophouse growers who are limited to the warm, high-light parts of the year may be able to start out and maintain their final density for the duration of their season.

An example of this in long-season eggplant growing would be transplanting single-stemmed plants at a density of 1.8/m^2 in January or February when the light intensity is low. Almost immediately after transplanting, the sucker below the first flower cluster on each plant can be developed into a second vine to double the density to 3.6 stems/m^2. Later, in good summer weather, a third leader can be developed on each plant for a density of 5.4 stems/m^2.

For tomatoes, an example might be planting 1.8 two-headed transplants/m^2 for a beginning density of 3.6/m^2. One of the reasons this works is that immature plants need less light than mature ones, so the plants mature as light levels increase. A month or two later, when light levels are getting really good, a third shoot on every other plant may be developed for a final planting density of 4.5 stems/m^2.

Lastly, it's worth noting that 3 feet (90 cm) is only 90 percent of a meter, so I should add the disclaimer that this method is really for figuring out the number of stems per 9 square feet. This has always been close enough for me, and it will give you a good idea of how close your plants are. Of course, if you want to be metrically correct, you can adjust to get the actual plants per square meter.

Vining/Fruiting-Crop Cycles

Most of the cropping cycles used for the vining/fruiting crops in protected culture are variations on a few different models.

Long Crop Cycles

Vining and fruiting crops take a long time to come into production; long crop cycles – which in heated production may be as much as a year long – therefore include long harvest periods to compensate for this fact. In more northern climates, tomatoes, peppers, or eggplant might be planted in January and grow until the following December, meaning the greenhouse might only be empty for a few weeks, just long enough to clean out the old crop before a new one is introduced. The cleanout is timed to coincide with the coldest, lowest-light times of the year. In this manner, growers as far north as southern Canada can keep plants growing for eleven-plus months without having to go to the expense of installing supplemental lighting. This is only possible because while the plants are seedlings and after they are topped (two months before they are removed from the greenhouse), they require less light. The season is timed so the plants' lowest light needs coincide with the darkest times of the year.

Because they no longer have to sustain vegetative growth, plants that are topped can ripen a fruit load that they otherwise wouldn't be able to support in the dark part of the year. Topping is the practice of cutting off the growing point (the head, or top) of the plant so that upward growth is stopped and all the energy goes to fruit production. Likewise, at the beginning of the next crop cycle, seedlings need less light to grow than do mature plants. So the new plants can grow at lower light levels than would support older plants. Growers in the north who want to grow a crop throughout the winter may have to install artificial lighting, depending on the amount of light they receive (see chapter 5).

Growers farther south may use crop cycles of the same length, but during opposite seasons. Instead of avoiding the coldest months of the year, many growers in southern areas find it difficult to keep greenhouses cool enough during the hottest part of the year. So growers might have an eleven-month season, with cleanout during the hottest month. In this way northern and southern growers can enjoy counter-seasonal production.

Short Crop Cycles

Tomatoes (especially determinate tomatoes), peppers, and eggplant can also be grown in one or two shorter crop cycles instead of one long one. Usually it's not economical to do more than two crop cycles per year with fruiting crops in a single structure, because too much time is taken up with establishing young plants that are not yet bearing fruit. This is the main disadvantage of shorter crop cycles: More of the season is devoted to growing new crops that aren't yielding. The advantage of shorter crop cycles, however, is that they are easier to manage than keeping a crop healthy and productive for nearly a whole year.

Regardless of geography, in a two-crop-per-year system it's most common to have a spring crop that yields into the heat of the summer, with a fall crop planted after the hottest part of the summer. Alternatively, sometimes growers will plan to remove an early-spring crop when their summer field tomatoes, peppers, or eggplant come in, and plan a fall crop to start maturing once field production drops off.

The equivalent of a short crop for cucumbers is the three- to four-month cycle preferred by most growers who use umbrella-style trellising. The longest a cucumber crop is normally grown is six months, using high-wire production (see chapter 10).

Pruning

Though commonly practiced on all the vining/fruiting crops in protected agriculture, the goals and strategies for pruning are often misunderstood. The best way to define pruning for protected agriculture is as a technique used to limit the amount and type of growth in order to keep plants balanced. Balanced plants will be more productive and healthier, with more predictable yields than those of unbalanced plants.

Plant balance is best understood by considering what it means for a plant to be unbalanced. All of the vining crops in this book form a sucker at each node. If left to its own devices, a plant would develop a new vine at every single leaf. Consider that every sucker would then begin producing a sucker itself at every node, and the number of heads would start to become problematic. If a plant were allowed to develop an unlimited number of heads, its finite amount of energy would be divided among so many heads and fruits that growth and ripening would slow considerably. I have seen this happen in indeterminate basket-weave tomato production, where ripening is noticeably slower and fruit size is smaller. With wide spacing this might not be as much of a problem, but when planted tightly in protected culture, unpruned indeterminate plants will branch out and compete with one another.

Pruning is one of the techniques that allows growers to crowd as many plants as possible into a finite area, by limiting each plant to the minimum amount of space it needs. Pruning not only conserves space by preventing vines from branching out, but it also makes a plant more open and permits light and air to flow through the canopy. This results in a minimal amount of yield reduction due to shading and disease. Fruit on pruned plants sizes up and ripens more quickly than on unpruned plants. And all crops that are kept pruned are easier to harvest, because you don't have to hunt through as much foliage to locate the fruit.

Pruning allows you to change the plant to balance it to its ideal size. For example, to grow a weak tomato variety over a long season, you might want to limit plants of that variety to only one vine, to focus all their energy up through the one stem, and hopefully keep it vigorous over the whole season. If you grew the same weak variety grafted to a vigorous rootstock, you might be able to grow with two heads instead of one.

Growing indeterminate tomatoes unpruned in a basket weave is usually only possible in arid climates; in humid areas, the foliage is so dense that moisture builds up inside the plants' canopy, which leaves them prone to foliar diseases.

Figure 6.6. This tomato leaf was improperly removed, leaving a stub, which may cause an infection in the stem.

Basics of Pruning

For specifics on how to prune each species, see the individual crop chapters. No matter what crop you are working on, though, it's always best to cut whatever part of the plant you're removing back to a living stem. Anytime you cut off a plant part and leave a stub, the stub may rot back into the stem and infect the rest of the plant.

Morning is a good time to prune, especially if you prune by snapping with your hands. Plants are more turgid in the morning and may snap more easily than in the afternoon, when they become more flexible. When pruning is done in the morning, plants have the

Figure 6.7. Botrytis moving into the plant through a pruning wound. Note the trellis clip that was correctly placed below the leaf before it was pruned off.

rest of the day for wounds to dry out, which will help them heal without infection. You can then spend the afternoon trellising, because the plants are more flexible when they are transpiring more, and easier to work on without breaking.

Pruning Suckers

Pruning goes hand in hand with trellising (see page 89) as a technique to get plants to grow the way you want them to. In addition to opening up the plant to facilitate light transmission and airflow, it keeps the amount of foliage and fruit balanced, which will keep plant growth and yield consistent. This may not sound like a huge issue, until you have a ton of fruit to sell one week but only half a ton the next. When you can't get rid of all of it the first week, and then don't have enough the second, you will find yourself wishing you had picked three-quarters of a ton both weeks instead.

Pruning Leaves

There are times when you will want to remove some mature leaves from most of the fruiting crops. With tomatoes and eggplant, this is regularly done below the fruit to increase airflow. Increasing airflow will reduce the

Figure 6.8. This tomato plant was not suckered on time, so the sucker is now as big as and competing with the main stem. You can tell the main stem is on the left because it has the flower cluster on it. The sucker to the right has grown out of the node of the leaf below the flower cluster.

amount of disease and help fruits to ripen more quickly by keeping them warmer. Older leaves can be removed without detriment to the plant because they are in the shade and don't contribute much to photosynthesis.

Pruning Flowering/Fruiting Clusters

All fruiting crops are prone to setting more fruit than they can mature. Pruning flowering or fruiting clusters is done with all fruiting crops at some point to put the amount of fruit set in balance with the amount of fruit the plant can actually support. See individual crop sections for fruit and flower pruning recommendations.

Trellising

Most protected fruiting-crop growers use a trellis to maximize the health and productivity of their vines. Ease of harvest and ergonomics are other important considerations. In the field, any of these crops are frequently grown without support, but they are almost always trellised in protected culture. Protected fruiting crops are the apartment dwellers of the vegetable world: By directing the plants' growth upward – taking advantage of the structure's height – instead of sideways, you can grow them at a higher density to make the most of precious real estate.

Consider a typical greenhouse eggplant planting density: A double row of eggplant is planted 2 feet (60 cm) apart, with plants in the row every foot (30 cm), separated by 3-foot (90 cm) paths. With no support at this spacing, the plants don't have to grow for very long before they are flopping all over one another and out into the pathways. And if the plants are flopping on one another, they are shading and inhibiting one another's growth. If light is blocked, so is airflow, which will make the development of diseases more likely.

Flopping also means the grower's view is blocked, so he or she will have to move the plants around to find the fruit, which slows down harvest. Fruit on the ground means more dirty and damaged fruit. Lastly, it is much more difficult, if not impossible, to prune flopping plants. Simply put, there's no way to grow vining crops at such high density without trellising.

I have occasionally seen growers plant a sprawling crop of cucumbers in a hoophouse or screen house. But they could get a lot more out of the same size structure, and get a lot more cucumber plants into it, if they put in a little more labor in the form of trellising. It is so much easier to harvest cucumbers that are held off the ground than to bend over and hunt through the leaves.

Trellising requires an up-front investment of time in order to make jobs down the road, like harvesting and pruning, go faster. It goes hand in hand with pruning to get the plants to grow the way you want them to. It's also an investment in plant health and clean, marketable fruit.

Structural Considerations for Trellising

When you're attaching trellising to your structure, first make sure the frame is in good shape and able to support the load of the crop plus whatever exterior forces may impact the structure, like snow and wind. Some greenhouse manufacturers offer reinforced end walls designed to take the extra force of trellising. They may also be able to provide information on how much of a load their structure will support.

All common greenhouse trellising methods use twine in some fashion to support the growing plants. The high humidity and extended season of the greenhouse environment tend to break down natural fibers like jute and sisal, reducing their reliable service life to less than one greenhouse season. Even if natural fibers don't break, they have a tendency to mold in humid growing conditions. Since it's impossible to predict what type of mold will develop, and it could potentially be a plant pathogen, many growers avoid using natural fibers for these reasons.

In the past, plastic was the only choice for growers who wanted to avoid natural fibers. Now compostable bioplastic options are available, which are not only greener than plastic, but also save labor. Disentangling plants from noncompostable twine and trellis clips takes a lot of time. Now with compostable twine and clips, being able to simply cut the plants down and compost everything at the end of the season is a huge time-saver.

Gripple

Originally designed to make installing fencing easier, the gripple is useful for installing trellis wires. It can be difficult to attach wires to solid objects (like greenhouse frames) while keeping them under tension, and then undo them without cutting when they need to be moved. The gripple solves both of these problems.

The gripple is a small piece of metal that you use in conjunction with a plier-like tool to join wires together and tension them at the same time. You put the gripple on loosely and then tighten it to your desired tension. It forms a very secure connection that you can also release if you need to move wires. There are also gripples for installing hanging wires, which could be used to install lights or anything else that might need to be hung from the structure. This method does require buying the installation tool and some gripples, but it makes setting up trellis wires much easier and faster than older methods like twisting wire or using ferrules.

Trellis Styles

A number of different styles of trellising are used in protected culture. The most common method involves growing each vine along support twine to keep it off the ground. A basket-weave-type trellis is sometimes used as well.

Basket Weave

The basket-weave style is borrowed from field growing and can be used with tomatoes, peppers, and eggplant. This method is good for growers who have structures that can't support the weight of trellised crops, or who are looking to save time spent trellising. Basket-weave trellising is sometimes called Spanish style in protected culture, since this is a common method for trellising in the tens of thousands of acres of low-tech tunnels in southern Spain.

To basket-weave, you plant a row of plants, pounding a stake between every few plants (a fairly common ratio is two plants, then a stake, repeated all the way down the row). In field growing, 4- or 5-foot (1.2–1.5 m) wooden stakes or T-posts are often used, sunk about 1 foot (30 cm) into the ground. Longer stakes are more often used in protected culture, where the longer season and lower light levels contribute to taller plants than in the field, even with relatively short varieties like determinate tomatoes. Next, run a piece of twine down each side of the row of stakes to contain the plants, adding another level of twine every time the plants grow 8 to 12 inches (20–30 cm), before they flop over. In this manner the "weave" contains the plants and keeps them growing vertically, but you don't have to spend time trellising each individual plant as in the other trellis methods. When this method is used with peppers, string is usually only applied on one side at a time; otherwise a lot of peppers get trapped between the strings. Fruits trapped between strings will be deformed and scarred from the pressure of the twine.

Basket weaving has to be done before the plants flop out of the trellis. If the plants have started to flop, it will take much longer to tuck them back into the trellis and apply string at the same time. A disadvantage of the basket weave is that the plants are difficult to prune and work on since they're trapped between the strings. This is why basket weaving is most commonly used with determinate tomato varieties that don't need pruning.

String Trellis

The most common way of trellising vining crops in protected culture is to attach a piece of twine to an overhead wire where each vine is to be trellised, and grow the vine up the string. I can't guarantee that this will hold under all circumstances, but 12-gauge high-tensile wire with a noncorrosive coating is used in many greenhouses to support the trellis twine. Some growers prefer aluminum aircraft cable; since it's made of multiple strands, it isn't prone to single-point failure like monofilament wire. Others attach trellis strings to pipes they've laid on the trusses of the structure over where the rows will go.

I would advise you to put a string over every plant, but you might have plants with multiple heads, so the ideas is to put a string above every vine. For example, in a greenhouse eggplant crop with a vine every foot (30 cm), you'd tie a string every foot along the overhead support wire.

Figure 6.9. The view from the canopy of a tomato greenhouse. Note how the tomahooks wound with blue twine are hooked on aircraft cable to facilitate lowering and leaning.

To support each individual plant, loosely tie or clip twine to the base of the plant using trellis clips. It's important that the knot be loose, because a knot snug to the plant's stem will cut into it as it grows. Tying is more economical, whereas using clips may be quicker. Using clips is better for inexperienced workers, since there is no chance for them to tie the knot too tight and damage the plants.

When it comes to supporting the vine as it grows, you can use clips, or twist the string around the vine as it gets longer. Once again, twisting is cheaper but clips are probably easier for inexperienced workers, and both methods are pretty fast. Twisting has a slight generative effect (see chapter 7) on plants when compared with clipping, but I think growers mostly choose a method based on personal preference. Still, I know some growers who twist plants they want to be more generative and clip plants they want to be more vegetative.

To use the twisting method, simply twist the stem of the plant around the string as it grows to keep it from flopping over. Once you start twisting in one direction, you have to keep twisting in that direction; reversing it

Figure 6.10. This trellis clip was installed properly under a leaf to help support the plant.

Figure 6.11. These peppers are being trellised simply by twisting the string around them as they grow. The strings are tied to high-tensile wire overhead.

will un-twine the plant and cause it to flop off the string. It's best to pick a direction and use that on all your plants, so you don't have to pause and assess each plant to see which way it's twisting.

If you've never done this before, it may be hard to imagine. But all you have to do is try it and see how it works. Make sure not to trap any flower clusters or other plant parts between the stem and the string. Twist the plants on a regular basis so they don't flop off.

To use clips for trellising, just add a clip as the plant grows, before it flops away from the string. Clips are available in plastic and bioplastic, which can be composted along with the vines at the end of the season. The technique is to steady the plant close to the string with one hand, and get a clip with your other hand. Clips are sold in the "open" position so they come ready to use. Bring the open clip around the stem, placing the twine in the hinge – this will hold the clip securely in place – and close the clip so it clicks. Place clips beneath sturdy, fully expanded leaves to help support the plant. Do not place clips below flower clusters, as they may interfere with fruit growth.

Don't work so high up on plants that you're clipping or twisting growth that isn't fully expanded. If you clip or twist the immature part of the plant that's still expanding, this will hinder the plant's upward growth. If you have any doubt about what fully expanded growth is, look down the plant and see what size the mature leaves are. Don't do any trellising above the mature leaves.

Figure 6.12. An example of an improperly installed trellis clip. It was put on the plant too early, before the plant had finished elongating. So as the plant extended upward, the flower cluster pushed against the clip, creating the tension visible between the two clips and deforming the plant's growth.

If the leaves aren't close to their mature size, that part of the plant may still be elongating; this will cause it to grow into the clip, deforming the plant (see figure 6.12). Be sure to clip plants before they flop away from the string, which may break or kink the plant. Even if it doesn't hurt the plant, it will take extra labor to clip a plant that has flopped over rather than clipping it once right by the string.

Whether you twist or use clips, it's important to trellis on a regular basis. If plants grow so much between being trellised that they flop, the vine may snap or the bend may cause damage to the stem. And it's much faster to trellis plants that are growing along the twine than it is to deal with plants that have flopped. For fast-growing crops like tomatoes and eggplant, trellising may need to be done on a weekly basis to keep up with the rate of plant growth. Slower-growing peppers may only need attention every other week. Fast-growing cucumbers may need attention even more than once a week.

SUPPORT WIRE LAYOUT

A standard way to space rows and overhead support wires is to form a double row by installing two wires 2 feet (60 cm) apart above the crop, separated by a 3-foot (90 cm) walkway. The nice thing about standardizing greenhouse/hoophouse layout with this trellising system is that all the common spacings for vining crops can be accommodated. This means you can grow any of the vining crops anywhere in the house without having to change trellis wires. Even if you decide a standard spacing is too dense for you, you can reduce density by increasing spacing between heads along the wire without having to move the wires. The same double-row layout can be used with basket-weaved crops.

Lowering and Leaning

Lowering and leaning is a variation on the string trellis that is very common in hoophouses and greenhouses growing long-season vining crops. It's an important protected culture technique because it allows you to grow a vining plant taller than your structure. You can grow a plant of almost unlimited length in the finite space of a greenhouse by lowering the head of the plant and moving it down the row, trailing the vine along behind it as it outgrows the available vertical space.

Lowering and leaning is the standard way of growing indeterminate greenhouse tomatoes. In the past, a minority of cucumbers were grown this way, but lowering and leaning cucumbers is becoming more popular. When cucumbers are lowered and leaned it's called the high-wire style of cucumber growing. Eggplant can be lowered and leaned, but it has to be done carefully because their stems are not as flexible as tomatoes' and cucumbers'. Peppers are not usually lowered because their stems are too brittle. Plus, they are slower growers than the other crops, so they don't use up their vertical space as quickly.

Several types of tools make lowering and leaning possible. Instead of tying twine to the overhead wire, growers make use of a tool that holds a spool of twine up by the wire. This spool clips to the wire and lets down enough string to reach the plant, holding the rest in reserve until you need it. When the plant outgrows its space, some string is let out (about a foot at a time), lowering the plant, and the head is moved down the wire so the stem doesn't bend sharply and snap.

How to Lower and Lean

Before you can lower and lean a crop, you need to do some planning. First, it helps to have a tall greenhouse. If you try to lower and lean with a short wire height – 6 feet (2 m), for example – there is a good chance the plant will grow to the top wire and need lowering before you start harvesting the fruit. This is bad both from a food safety standpoint (you don't want the fruit touching the floor), and from a marketability standpoint (fruit that makes contact with the floor will get scratched up and dirty and become less marketable).

In addition to attaching the plants to the overhead wire with something that can be easily removed, you need to arrange them in double rows. This is because the plants on the end of each row need to be looped around to the other side, to make space for the plants behind them to lean.

To begin lowering and leaning on an end, take the last plant in a row and lower it. Once there is slack in the stem of the end plant, loop it around to the end position of the opposite row. This plant will temporarily be overcrowded, until you finish lowering and leaning. To help the plants make a gradual turn around the end of

Figure 6.13. You can see how these tomatoes are being lowered and leaned, with one side of the double row going in one direction and the other side in the opposite direction. The plants are growing in 5-gallon buckets.

the row, many soil growers drive a post to help the plant make a nice loop around the end.

In many greenhouses, lowering and leaning is a weekly task, so to stay ahead of the plants you will need to let out enough twine to equal a week's growth. In most tomato crops, this equals two spools off a tomahook, or two turns of the spool of a roller-type hook. Cucumber plants that tend to grow faster than tomatoes will need more slack to be released, or they'll have to be lowered more than once a week.

With some tools for lowering, like the tomahook, you have to take the spool and manually rotate it to let some twine off the hook. There are others, like the Rollerhook or RollerPlast, where you push a button or use another mechanism to release twine from the spool.

Once you've lowered the plant, the stem will be slack and may have a sharp curve at the bottom. This may not seem like a big problem, but if it happens week after week the stem will eventually kink and potentially snap. In order to keep enough tension on the vine that it doesn't kink, you need to move the plant in the direction it will be leaned as you lower it. This is why you begin the process of lowering and leaning by moving the end plant around to the other row. You move on to the second plant and, after lowering it, move it into the space created by moving the end plant over to the other row. Lowering and leaning is sometimes called racetracking the plants, because they end up circling around the ends in a long oval, the way cars drive around a racetrack.

So every time you lower and lean a crop, the end plant on each row moves over to the adjacent row, and the plant that was second from the end moves into the end position. When you get to the end of the first row, there will be an empty space created by moving all the plants down the row in the opposite direction. This is where you will put the end plant from the opposite row after it has been lowered and looped around the end. This creates an opening on the end of the other row into which you can move the next plant, and on down the row. When you get to the end of *that* row, you will come to the very first plant you moved, the end plant from the first row, which is overcrowded. As you move the next-to-last plant on the second row, this will restore the proper spacing and you can move on to the next row.

Figure 6.14. This grower has installed a ridged pipe at the end of his rows to help the tomato vines make a gradual turn without snapping when they are lowered and leaned.

Lifts and Other Ways to Work on Tall Plants

Since lowering and leaning does not work well in low structures, lowering the plants cannot usually be done by someone standing on the ground. You can use a variety of methods to reach the tops of your plants. In smaller structures, you may be able to use ladders, kick stools, or other stationary methods to reach the wire. These stationary methods are time consuming, because you have to get off and move them every time you want to move down the row.

There are also a variety of carts that can be purchased or made. Many of the commercial models are designed with wheels that run on pipe rail, so they don't need steering. This makes for a very efficient process in large greenhouses, as they are controlled by foot pedal so the person doing the lowering and leaning can keep the cart moving with both hands free. Many greenhouses have scissors lifts that can be adjusted to work at any level. Cheaper versions work more like a scaffold wagon, with platforms for people to stand on that have to be set manually.

These lift carts are also useful for working on and maintaining the roof and trusses of a structure. I have also seen a variety of homemade and nonmotorized carts that growers have made themselves. The challenge with carts that don't run on pipe rail is steering – you don't want the carts to run into the crop rows.

Post-Harvest/Storage

The way a crop is handled after it is harvested has a lot to do with how long the produce will store. The individual crop chapters describe the best post-harvest handling practices to maximize shelf life and quality.

One of the nice things about growing fruiting crops in a greenhouse is how clean they are. Since they're never in contact with the ground, they should never need washing. This creates efficiency during harvest since you can pick fruits straight into the containers in which you'll sell them.

Each crop has a different temperature and humidity level that needs to be adhered to in order to maximize the shelf life, quality, and flavor of the produce. Another important consideration is whether the produce is damaged by ethylene. Ethylene is a gas naturally given off by ripe produce, like tomatoes and apples. Some produce is very sensitive even to small amounts of ethylene. If exposed, their ripening and rotting process will accelerate, which may render sensitive produce unsalable in a short period of time. Take the temperature, humidity, and ethylene sensitivities of the crops you want to store into consideration when deciding how many separate produce storage areas you might need, since not all crops are compatible if stored together.

CHAPTER SEVEN

Temperature Control and Crop Steering

Temperature is one of the biggest factors determining both how a plant grows and the conditions of the growing environment. There are several approaches to managing the temperature of your greenhouse or hoophouse. On the simpler end, temperature can be set within a range that plants like to keep them happy and active. This is important since when plants go outside their comfort range, they aren't growing. For many growers, it's enough to keep their plants within a certain range and know that they aren't experiencing temperature stress.

However, environmental control can be used to do a lot more than just keep plants within a certain happy range. For example, temperature can be used to make the plant more or less vegetative, and to make the environment wetter or drier if you know how to manipulate it.

Our understanding of climate control has come a long way over the last few decades. Computers that can control heating and venting systems, and instrumentation that can measure environmental conditions, have allowed researchers to study and understand how plants respond to a wide range of environmental stimuli. Whether you're going to use more complicated climate control systems or keep it simple, it's important to understand the ways you can use climate to manipulate crops.

Active Plants

The active plant is an important concept in greenhouse growing. An active plant is one that performs the basic functions of respiration, transpiration, and photosynthesis, in contrast with a plant that has shut down due to adverse environmental conditions. This is where

Temperature and the Stage of the Crop

The fruiting crops have different ideal temperature ranges depending on the species and stage. Unlike leafy crops, they require you to balance vegetative and generative (fruit) growth at the same time. A smaller day/night differential at the beginning of the crop cycle will encourage vegetative growth and rapid establishment. Later, when the plants start flowering, a wider temperature differential should be used to promote fruiting. Temperature ranges for each crop are found in the individual crop chapters.

knowing the ideal conditions for each crop is important. Since you can control the conditions in protected culture, it's your job to keep them in the range where the plants stay active.

If a plant isn't active, it isn't growing; if it isn't growing, it isn't producing anything for you, though pests and diseases can still attack it. So any amount of time that a plant spends inactive reduces yield and leaves it more susceptible to damage. The goal is to keep plants active at all times, even when it seems like they aren't doing anything. Respiration should always be taking place, even at night. If it gets too cold at night, the plant will shut down and not benefit from respiration. When plants shut down at night it also delays them from becoming active again in the morning until they have warmed back up, which can take time.

One way to understand the effect of temperature stress on your plants is to think of your own reaction to extreme temperatures. If you found yourself in an environment that was way too hot or cold, you probably wouldn't be very productive, either. You'd spend most of your energy trying to moderate your temperature, by fanning yourself in the heat or doing jumping jacks in the cold.

Plants can't warm themselves, so they just shut down when it gets too cold. However, the reaction to excessive heat in people and plants is more similar. As the temperature increases, so does transpiration as the plant tries to cool itself. Evaporation is endothermic; it disperses heat. So transpiring the plant's water into the air is a way for the plant to cool itself. If the temperature gets so high that all the water absorbed by the plant needs to be used for cooling, growth slows until water becomes available for something besides cooling. And if a plant's need for water outstrips the supply, stomata may close in an effort to conserve water before wilting occurs.

If you touch the leaf of an active plant in a greenhouse, it should feel cooler than the air temperature. If you touch a leaf and it feels hot, it probably means that the plant's stomata are closed and it's about to wilt. If your plants' leaves are hot, this may be an indication that your irrigation system isn't functioning properly, or that you should provide some extra water on a hot day.

The rates of respiration and photosynthesis both rise with the temperature, though the rate of photosynthesis levels off at some point. The amount that photosynthesis exceeds respiration is known as the growth potential. In tomatoes, for example, at around 96°F (36°C) the rate of respiration exceeds the rate at which assimilates are manufactured through photosynthesis. Since at this temperature a tomato plant is not making a surplus, and is probably experiencing an energy deficit, you want to avoid this as much as possible.

Heating for Warmth Versus Heating for an Active Climate

The simplest way to control the temperature in a greenhouse would be to set a low point on the thermostat and walk away, satisfied that the crop will not go below the set temperature. The crop will grow faster, be more productive, and be healthier than the same crop in an unheated house. The main benefit of a climate control strategy like this is that it is simple to manage and operate.

In a greenhouse without automated vents, all you have to do is pick a set point for your heater and decide when to roll your sides up. If you do have automated vents, it can be as simple as setting a low temperature for the heater to come on and a higher temperature for the vents to open. This is a very workable situation; many growers maintain greenhouses this way. However, greenhouses can be made more productive with more complex climate management goals. Below are two strategies that go beyond minimum/maximum temperature by considering more of the factors that affect plant growth.

Making Gradual Changes to the Growing Environment

Just like you, plants appreciate gradual changes in their environment. Think about how unpleasant it is to go outside from a warm house on a freezing day without a coat. This principle can be applied generally to managing plants in protected culture. If you want to make a change to the growing environment, do so in a gradual manner rather than a sudden one.

When it comes to heating, for example, a good rule of thumb is that when you need to change the temperature, don't make changes more rapidly than 2°F

(1°C) per hour if possible. The easiest way to do this is to decide what temperature you want it to be at what time, and count backward to determine when to start changing the temperature. Let's say you want it to be 68°F (20°C) when the sun comes up at seven o'clock in the morning. If your night temperature is 62°F (17°C), you will want to give the greenhouse three hours to warm up the 6°F (3°C), so start raising the temperature at 4 AM. This can be called a pre-day treatment. This strategy is used instead of setting a nighttime low and letting the sun raise the temperature, and has several big advantages.

When the sun does the heating, the temperature tends to spike. As soon as the sun hits the structure, temperature rise due to the greenhouse effect is very rapid. This is great for getting the temperature up but shocking to the plants, giving them just a few minutes to adjust from nighttime to daytime temperature.

Rapid warming in a humid environment also results in condensation. As the air temperature increases, moisture in the air condenses on the plants, which are colder because they don't change temperature as quickly as the air. Since wetness is a precondition for many plant diseases, letting the plants get wet makes them more likely to become diseased and nullifies one of the biggest advantages for a greenhouse grower over field growing.

Raising the temperature before the sun comes up in a pre-day treatment means that as the air warms up, the relative humidity will go down. With lower humidity, plants will be able to transpire more than they do in cool air that's saturated with moisture. This is important because if plants are unable to transpire, they can't grow. Raising the temperature before sunrise helps the plants transpire and be active as soon as the sun hits them.

Fruiting-crop growers may shoot to be at the desired daytime temperature even earlier, a half hour or hour before sunrise. If you get the temperature up to the desired level just in time for sunrise, the fruits may still be cold since they change temperature more slowly than both the air and leaves, and if the fruits are cold they aren't active and can't start taking assimilates from the leaves. So if you want the whole plant to be active as soon as photosynthesis starts in a fruiting crop, raise the temperature earlier to warm up the fruits.

Heating to Reduce Humidity

Power venting is what some growers call setting the ventilation point lower than the heating set point. This is a counterintuitive scenario where the heat is on at the same time the vents are open. Though it might seem like you would never want this situation, it's actually a very useful way to reduce the amount of humidity in the greenhouse.

Power venting works because it expels warm, humid air through the vents, and draws in and heats cooler air. The cooler outside air cannot hold as much moisture and thus is drier than the inside air. As the cooler air warms up, the RH goes down, VPD goes up, and it gains capacity to take on moisture.

Keeping the humidity down has two big advantages. For one thing, condensation is reduced, which will keep the plants healthier. The other advantage is its positive effect on respiration and transpiration, since as humidity increases, respiration becomes more difficult until plants can't give off any more moisture through transpiration. When plants can't transpire, they shut down until the humidity goes down.

Two important factors to consider when determining when and how much to power vent are condensation and humidity. Prevent condensation by keeping the temperature above the dew point, as this represents the intersection between temperature and humidity where condensation forms. This must be done proactively to be effective. If the plants are already wet when you walk into the greenhouse, it's too late; the goal is to prevent condensation in the first place.

Power venting can be useful early on cool mornings, when the humidity is high from overnight transpiration but low temperatures and lack of solar radiation are preventing the vents from opening. The intuitive approach would be to keep the vents closed tight to avoid using any more heat than necessary. But to maintain an active climate, set the ventilation a degree or two below the heater for just long enough to get rid of excess humidity. The extra money spent on heat to keep the air dry and the crop active via power venting more than pays for itself in improved crop productivity and health.

Condensation

One of the biggest advantages a greenhouse grower has is the ability to keep plants dry. This includes both keeping rain off and minimizing condensation. Moisture on leaves is necessary for the development of late blight and contributes to other diseases such as botrytis. Repeated cycles of condensation and drying are one of the contributing factors to the development of powdery mildew.

Condensation is temperature- and humidity-dependent. It occurs when air at a given humidity reaches its saturation point, known as the dew point, where the air cannot absorb any more moisture and water begins to condense on surfaces. Every protected culture structure creates its own mini climate and tends to be higher in humidity than the surrounding environment due to the transpiration of the plants releasing moisture into the air.

As evening approaches, the environmental conditions inside a protected structure are similar to those that lead to the formation of rain clouds outside. Clouds form when a humid air mass cools as it rises, with its moisture-holding capacity diminished the more it cools. When the air doesn't have enough capacity to hold its moisture anymore, the moisture becomes visible as a cloud. As the cloud continues to cool and lose more water-holding capacity, the moisture may begin to precipitate out as rain.

An unheated hoophouse at the end of the day contains an air mass humid with the day's transpiration. As the sun goes down and the temperature drops, the humid air mass loses water-holding capacity, just like a rising cloud. In a hoophouse the moisture will condense on plants rather than falling as rain. Greenhouse growers can prevent this by heating the air to maintain its water-holding capacity, or power venting to expel moisture.

If it doesn't happen overnight, condensation is likely to occur in the morning as the air warms and the plants stay cold. Hoophouse growers can't do much about this, but it may help to roll up the sides and open the vents on a hoophouse before dawn, to moderate the increase in temperature and release humid air. The ability to control condensation is one of the big advantages of a greenhouse over a hoophouse, and well worth the cost of the fuel to heat and prevent it.

Crop Steering for Generative or Vegetative Growth

Though I didn't know it at the time, one of my earliest gardening experiences was with tomato plants that were excessively vegetative. My family had an enormous compost pile in the same place for many years. When we finally decided to put a garden in, I thought it was going to be great with all that compost. We smoothed out the heap, rototilled it, and the soil looked like chocolate cake.

We put tomato plants in and they started growing like Jack's Beanstalk, but they never set much fruit. And the fruit they did set was smaller than it was supposed to be. Like any overeager gardener, I went and checked the fruit every day, and they took forever to ripen. It wasn't my constant checking that made it seem like they were taking so long – the days to maturity listed in the seed catalog confirmed they really were ripening slowly. I didn't know it at the time, but I had excessively vegetative tomato plants.

When a plant puts more of its energy into leaves, stems, shoots, and roots, it's in a vegetative state of growth. If it puts more of its energy into flowers and fruit, that's called a generative state of growth. When a plant puts an adequate amount of energy into both at the same time, we say it is balanced.

As a fruiting-crop grower, most of the time you want your plants to be in a balanced state of growth. This is in contrast with leafy-crop growers, who want to keep their plants vegetative. When leafy crops start putting energy into generative growth, it causes the development of flowers and seeds (bolting) that make the crop unmarketable. But there may be times when unbalanced growth is desirable. For example, right after transplanting you want the plant to put its energy into vegetative growth to develop a strong root system as quickly as possible. And at the end of the season, the plant might want to put all of its energy into generative growth to make as many fruits and seeds as possible before a killing frost.

There are a lot of environmental factors that influence whether plants are vegetative or generative. With control over the environment in protected growing comes the ability to manipulate, or steer, plants in the direction you want them to go. Balancing fruiting and vegetative growth over a long season is one factor that makes growing fruiting crops more complicated than growing leafy crops. Recognizing the state of growth the plant is in and having the ability to change it when necessary is one of the big advantages you gain in protected growing.

When my family got the results from a soil test at the end of our first gardening season, the fertility and organic matter levels were off the charts. In our inexperience, we didn't realize that you could have too much of a good thing (fertility), or that humble compost could result in overfertility. If you want high, consistent yields of fruiting crops that ripen in a timely manner over a long season, you want plants that are balanced.

Once you learn the signs of a plant that is balanced or unbalanced, you won't be able to miss them. Which is why I am grateful for my experience in the garden for teaching me what an excessively vegetative plant looks like. It's important for fruiting-crop growers to be able to assess and understand the state of the plant: A grower who can interpret the signs can tell where that plant's energy is going, and whether it needs to be changed.

Remember that the terms *vegetative* and *generative* are not absolutes. A plant does not switch back and forth like a light going on and off. *Vegetative* and *generative* are relative terms we use to describe states on the opposite sides of a continuum.

Partitioning

Partitioning refers to how a plant divides its resources between fruiting and vegetative growth. A plant has a finite amount of resources at its disposal. You want it to put some energy into the fruit so you have something to pick today, and at the same time to put energy into the plant so it will keep growing and there will be something to pick months from now.

It may help to think of each plant as two separate sets of parts: vegetative (roots, shoots, and leaves) and generative (fruit and flowers). Plants don't always put equal amounts of energy into both types of growth. Depending on many factors we will discuss below, a plant may prioritize one type of growth over the other. Key to understanding why partitioning matters is knowing that a plant may put so much energy into the fruit that it gets weak and peters out, or so much energy into the plant that there is very little fruit.

As the grower, it's your job to keep the plant healthy, so it has as much energy as possible. But you are not going to get high yields if you're not optimizing both the amount of energy the plant has and how it uses that energy. Creating a balanced plant – one that puts adequate energy into both fruiting and vegetative growth – is what will result in the highest yield and fruit quality.

Vegetative Growth

A plant is in a vegetative state when it puts an excess of energy into the plant at the expense of the fruit. Signs that a plant is vegetative include fruits that are smaller than they are supposed to be, fruits that take longer than normal to ripen, and a low ratio of fruits to leaves.

An excessively vegetative plant will have larger, longer leaves than normal. The stem below the head of the plant will be very large, and in tomatoes it may have a characteristic oval shape instead of being round. Flower clusters tend to be poorly formed and stick straight up in the air instead of curving out from the plant. These are

Table 7.1. Signs That a Plant Is in a Vegetative or Generative State of Growth

Criteria	Vegetative	Generative
Leaves	Long, thick, leathery, dark green	Short, thin, light green
Stem	Thick	Thin
Head of plant	Dense and bushy, with twisting leaves	Sparse and open
Flower trusses	Long, thin, sticking up straight, poorly formed	Short, thick, curled, well proportioned
Ripening	Slow	Fast
Flowers and fruit	Small, poorly shaped	Large, well formed
Flower color (tomatoes)	Lighter yellow	Darker yellow
Position of highest open flower on the plant	Far below the tip of the plant	At or close to the tip of the plant

called stick trusses, and they're in danger of kinking or snapping off under the weight of the fruit they hold.

Plants are particularly prone to becoming excessively vegetative at the beginning of the season, because there is no fruit load to balance the plant's vegetative growth. Plus it's the plant's natural inclination to grow very fast at the beginning of the season in order to become established as quickly as possible.

The problem of excessively vegetative plants may be particularly acute for hoophouse growers, who are limited by the weather as to how early they can plant. By the time they can transplant, the days tend to be long, with lots of light and heat. All these factors push the plant to grow quickly and vegetatively, which can have a bad effect on yield of valuable early fruit.

One time we grew a medium-sized greenhouse tomato called Massada, which was prone to going vegetative under our particular conditions. Of all the varieties we were growing simultaneously in our hoophouse, it was by far the most vegetative. As the other varieties with similar days to maturity started ripening, the Massadas had no color, and were smaller than they were supposed to be, topping out at 4 to 5 ounces (113–142 g) instead of 6 to 8 (170–227 g).

Days turned into weeks, and I kept visiting the plants every day to see when one would ripen, which only made the situation seem worse. When they finally ripened, the tomatoes were so undersized there was no way we could sell them loose as small beefsteaks as we intended.

Figure 7.1. This long, skinny flower cluster is known as a stick truss because of the way it sticks straight out from the plant. It's likely to bend or break under the weight of the fruit.

Figure 7.2. The downward-curling leaves and thick stems are an indicator that this tomato plant is excessively vegetative, or bullish.

Figure 7.3. This long, thin, poorly formed flower cluster is an indicator that this tomato plant is excessively vegetative.

Figure 7.4. This sucker developing off a leaf is an indication that the tomato plant is too vegetative. Some varieties are more prone to this than others.

Figure 7.5. The curling, twisting leaves of this tomato plant are an indication that it is excessively vegetative. Note that the open blossoms are well down below the head of the plant—yet another indicator that vegetative growth is outpacing generative.

Instead, we clipped the entire truss off as they ripened and sold them as cluster tomatoes, since that's about the size they were. Over the course of the season the plants settled down and became more balanced. They began putting more energy into the fruits, and they started sizing up and ripening on time. But the first few clusters on those plants were a disaster from a yield and income perspective.

Overly vegetative plants, which produce rampant leafy growth at the expense of fruit growth, are especially bad early in the season because they reduce the early yield. A lot of growers get their best prices from early fruit, so small, late fruit at the beginning of the season is doubly damaging to the bottom line. Excessively leafy plants are also bad for labor efficiency since a leafy, fast-growing plant with undersized fruit will require more pruning than a balanced plant with normal fruit size and production.

Generative Growth

A plant is in a generative state of growth when it puts more of its energy into fruit at the expense of the rest of the vine. A plant that is too generative might yield fine in the present, but production will fall off over time as plant growth slows.

Most of the signs of a generative plant are opposites of the signs that a plant is vegetative. Plants that are too generative will have a very thin stem at the head. In tomatoes, flowers will be a darker yellow than in vegetative plants. Trusses will be shorter, nicely proportioned and arching out from the plant instead of sticking straight up in the air. This type of stem architecture will support a fruit load much better than a stick truss.

In all species, generative plants will have larger, stronger flowers. Leaves will be smaller, flatter, and less curled than in very vegetative plants. Flowers will open closer to the tip of the plant. In a vegetative plant, the tip of the vine will grow faster than the flowers. But when a plant prioritizes generative growth, the flowers will open close to the tip because they are growing as fast as or faster than the vine. Fruit on generative plants tends to be larger, better proportioned, and quicker to ripen than on plants that are too vegetative.

Balanced and Unbalanced Growth

Unbalanced growth, regardless of whether it's too vegetative or too generative, is not desirable in the long run. Unbalanced vegetative growth would be great if you could sell tomato leaves. But since you can't, all that excessive leafy growth comes at the expense of fruit production. Excessively rapid vegetative growth also produces lush, soft plant tissue, which makes the plant more susceptible to injury and attack by pathogens. With unbalanced

Figure 7.6. The thin stem of this tomato, with blossoms open at the same level as the head, is an indicator that this plant is very generative.

Figure 7.7. The fact that this plant has open flowers so close (3 inches/7.5 cm) below the tip of the plant means that it's still on the generative side.

generative growth, the flowers and fruit may look great while the plant wastes away. Typical of a generative plant is big fruit with a small, wispy head that doesn't grow very fast. When a plant stays generative, yield goes down as plant growth slows and becomes weak.

One of the only times I can think of when unbalanced growth is desirable is during the first week or so after transplanting, when it's common to encourage vegetative growth in order to speed rooting and establishment of the transplants. This is where the additional control of a greenhouse comes in handy. The main way to encourage vegetative growth is to run a relatively high temperature for the species, with little or no day/nighttime temperature variation. This is known as a flat temperature profile. The high temperature causes the crop to grow quickly, and the flat temperature promotes vegetative growth. For a detailed explanation of temperature recommendations by species, see each individual crop's growing recommendations.

Creating balanced growth really is a balancing act in that crop conditions are always changing, and it's not a simple formula. You may have just gotten your crop balanced after transplanting for the spring season when the heat of summer hits and throws everything out of balance again. Balancing plant growth is a season-long job that's never really over until the end of the crop, which is why it's so important for growers to be able to read the state of the crop and know just by looking at it what type of growth it's experiencing.

Crop registration is useful for helping confirm trends that seem apparent in the crop. They can tell you whether or not the trend matches the way the crop appears. For example, if you thought your crop looked vegetative, and you had records from registration to refer to, an accelerating growth rate and thickening head would seem to confirm that the plants are trending vegetative. With this information you could use some generative actions to steer the crop back toward balance.

Sources and Sinks

Another way to think about balanced plant growth is in terms of sources and sinks. A source is where energy is generated, and a sink is where it is absorbed. The main sources of energy for a plant are mature leaves. The main sinks are fruit and immature leaves at the head of the plant.

The numbers of sources and sinks need to be balanced in order to have a balanced plant. As a grower, you have two jobs relating to plant balance. With steering, you need to make sure the plant is putting energy into both fruit and leaves at the same time. Your other job is to make sure that the plant's sources and sinks are proportional, so there are enough sources to feed the sinks, and enough sinks to absorb the energy generated by the sources.

An imbalance between the numbers of sources and sinks commonly occurs right at the beginning of crops that tend to be vigorous and vegetative, like tomatoes or eggplant. As soon as they are transplanted into the production house they begin vigorously growing leaves. A plant must direct its energy toward something, and if there aren't any fruit, it's going to put all that energy into plant growth. Because it's the beginning of the season and there isn't any fruit yet, all of a plant's energy is directed back into more leaf growth. This is why it's so common for young tomato and eggplant crops to be excessively vegetative. The fruit load on a mature plant actually helps to balance it out.

After a period of poor fruit set, when opportunities to set fruit have been missed due to high temperatures or poor pollination, is another time when a crop is prone to becoming excessively vegetative – putting all of the energy that would otherwise go into fruit into growing leaves. In addition to fixing the problems causing poor fruit set, it would be important to recognize the plants' bad ratio of sources to sinks, to be on the lookout for signs that they're becoming too vegetative, and to be ready to steer the plants back into balance with the actions described below.

Pruning too many leaves in relation to fruit load can leave plants with too many sinks in relation to sources. Foliar diseases that kill leaves or compromise their photosynthetic capacity can have the same effect, leaving the plant with more fruit than the number of leaves can support. In cases where there are too many sinks, you can reduce the fruit load by harvesting more aggressively or even removing an immature fruit or a cluster to reduce the demand on the plant. Sometimes rebalancing the plant in the long run is worth the short-term reduction in yield. If plants are struggling for any reason, lightening the fruit load is always an option to help them through a rough patch.

If growing were easy, plants would have meters on them telling you whether they were vegetative or generative, and dials to change the settings. In the absence of meters, you have to learn to read the signs. And in the absence of dials, you have to learn how to manipulate the environment to get the plants to grow the way you want them to.

There are a number of different factors a grower can manipulate in order to steer a crop in the desired direction. Actions that encourage vegetative growth are known as vegetative actions, and those that encourage generative growth are generative actions.

In addition to the influence environmental factors have on the state of crop growth, individual varieties bring their own influences as well. For example, some varieties are naturally more or less generative or more easily balanced than others. Keep this in mind, and pick varieties based on your growing environment. If the environment you are growing in is naturally vegetative, a generative variety might be easier to balance.

How Crop Steering Works

Broadly speaking, anything that causes the plants stress will function as a generative action, and anything that reduces the amount of stress on the plants is a vegetative action. Things that stress the plants make them think they're a little closer to death, so they focus on reproduction to pass their genes along.

There are a number of different ways a crop can be manipulated to steer it in the desired direction. Since many of the actions involve changing environmental conditions, they are mostly off limits for field growers.

Manipulating the Day/Night Temperature Differential

One of the most common ways to steer a crop is to manipulate the amount of difference between the day and night temperature. Growers call this the day/night differential, or dif for short. For example, a flat temperature profile with no difference between day and night temperatures would be the most vegetative. The larger the difference between the day and night temperatures, the more generative an influence it will be.

This works because when the differential is small, the growing conditions are more like summer, with warm days and nights, so the plants think they have a long season to continue growing and put lots of energy into vegetative growth instead of reproduction. But when the nights are much cooler than the days, the plants think winter is coming and they put their energy into reproduction.

Grafting

Most commercial rootstocks will influence grafted plants to be more vegetative. This can be desirable because it will allow a crop to thrive over a longer season and be more productive. The potential drawback (as with any very vigorous variety) is that the grafted plants may be difficult to balance early in the season due to their being so vegetative. Newer rootstock varieties have been developed to address this problem. More balanced rootstocks with less vigor may be worth trying, especially with naturally vegetative plant types, like small-fruited tomatoes and eggplant.

Most grafted plants have at least two heads to try to deal with the extra vigor. Single-leader grafted plants tend to be more difficult to balance. For more information, see chapter 8.

Pre-Night Treatments

Pre-night treatments are the closest thing to having a button that will send energy straight to the fruit. This technique can be used on all the fruiting crops described in this book, though the details vary somewhat with the crop. See the individual crop chapters for specific recommendations.

Referred to as pre-nights for short, pre-night treatments are a technique used to intensify the generative effect of the day/night temperature differential.

Table 7.2. Factors Influencing How Vegetative or Generative Plants Are for Crop Steering Purposes

Factor	More Vegetative	More Generative
Age of plants	Younger	Older
Twenty-four-hour temperature average	Lower	Higher
Night temperature	Lower	Higher
Difference between day and night temperatures	Smaller	Larger
Speed of change from day to night temperature (part of pre-night treatment)	Slower	Faster
Relative humidity	Higher	Lower
Amount of ventilation/air movement	Less	More
Water stress	Lower	Higher
Frequency of irrigations	Higher	Lower
Time of first irrigation of the day	Earlier	Later
Time of last irrigation of the day	Later	Earlier
Level of electrical conductivity (EC) in root zone (saltier is more stressful/more generative)	Lower	Higher
Stressful growing conditions	Less stress	More stress
Amount of nitrogen	Higher	Lower
Fruit load	Lower	Higher
Level of carbon dioxide in greenhouse	Lower	Higher
Amount of leaf pruning	Less	More
Taking a leaf out of the head (tomatoes)		Generative
Grafting (tends to make plants more vegetative, how much depends on rootstock variety)	Grafted	Ungrafted
Pre-night treatment—combination of three factors: amount, speed, and duration of temperature drop at end of day	Less, slower, shorter	More, faster, longer

Pre-nights are one of the few times you deliberately want to break the rule of making temperature changes gradually at a rate of 2°F (1°C) per hour.

The basic concept of a pre-night treatment is very simple. At the end of the day, when the temperature is going to drop naturally, keep the vents open and let it drop as rapidly as possible until the nighttime low is reached. This is instead of closing vents to conserve the heat of the day for a gradual decline to the nighttime temperature.

Pre-nights work because as the plants rapidly cool down, the fruits stay warmer than the other parts. Since water temperature changes more slowly than air temperature, the slowest parts of the plant to cool are those with the most water: the fruits. Pre-nights take advantage of the fact that parts of the plant that stay warmer continue drawing on the assimilates produced during the day longer than those that cool down first. Pre-night treatments are a way to direct the energy from the day's photosynthesis to the fruit. As long as there is an excess of assimilate produced, this can result in higher yields and better flavor, since you are drawing sugars to the fruit.

Pre-nights can also help with fruit setting in a crop that has had issues with blossoms falling off without forming fruit. Since peppers are especially prone to

Mixed-Crop Greenhouses

Each crop has its own environmental and temperature preferences. Many large greenhouses grow just one species at a time in order to tailor conditions to the needs of a single crop.

However, for many growers there is value in diversity. Whether having a variety of produce is a way to stock a farm stand, fill a CSA box, or appeal to produce buyers, many growers will want to grow more than one crop in a single greenhouse.

For successful mixed cropping, it is important to take crop compatibility into consideration. For example, vining/fruiting crops and lettuce/greens are not the best combination in a greenhouse. The height of the vining crops tends to shade the lower-growing greens crops when they are grown anywhere except to the south of the vines. The fruiting crops also prefer a hotter, more variable climate than lettuce would like. Lastly, the fruiting crops appreciate a taller structure than greens need. When the two types of crops are grown together either the height is wasted on the greens or the vines grow in shorter-than-ideal conditions.

Despite the compromises, lots of growers make it work. Take the needs of each crop into consideration, and make the compromises as amicable as possible for each one. Or you could pick two or more crops that have similar needs, an example being tomatoes and eggplant. Both crops have similar preferred temperature profiles and appreciate a pre-night treatment, so they can be grown together with little compromise in the production of either. Lettuce and greens grow well together, since they like cooler conditions and have similar nutrient needs.

Another good way to grow blocks of different crops without compromise is to put a curtain between sections of greenhouse to create separate climate zones. A thermostat can be used to control each section separately, or environmental control systems can be set up to manage more than one zone. This can be a good solution if you want to grow blocks of a few different crops with conflicting needs.

fruit-set problems, it's common for pepper growers to use pre-nights regularly to improve fruit set. It's also common for growers to use regular pre-nights on crops that tend to be vegetative, such as tomatoes and eggplant. Pre-nights are not as important in cucumber-only structures, since cukes aren't as prone to becoming excessively vegetative as tomatoes and eggplant, and they're not as prone to fruit-set problems as peppers. But if necessary, pre-nights can be used to steer cukes as well. This is where mixed-crop greenhouses can get tricky – if one crop needs a pre-night and another doesn't, the grower needs to decide which crop to favor.

If you are doing pre-night treatments and the crop still needs more generative steering, it is possible to intensify the effects of a pre-night treatment by widening the temperature spread. To produce a bigger temperature drop, some growers let the temperature ramp up a few degrees above the regular daytime setting right at the end of the day before a pre-night treatment. And then they drop it a few degrees below the average nighttime temperature right at the beginning of the night to really get a wide spread. Dropping the temperature extra low at the beginning of the night can also help reduce the average temperature when excessive daytime heat has pushed the daily average too high. Prolonging the amount of time spent at the pre-night temperature low is another way of intensifying a pre-night treatment, and potentially of bringing down hotter-than-ideal twenty-four-hour temperature averages.

Note that pre-night treatments can only be done on sunny days with lots of photosynthesis and a surplus of assimilates, since the plant has to generate a certain amount of assimilates just to keep itself alive and support continued growth. If you do pre-nights on cloudy days, too much energy may go into the fruit and weaken the plant, which may have generated barely enough energy to stay alive. And if you do pre-nights during a patch of overcast weather with multiple cloudy days in a row, the heads can thin out very quickly and the plants lose vigor. Since pre-nights will have a generative effect, you don't want to use them if the crop is already too generative.

The conditional effects of pre-nights are an example of how you can't just be on autopilot and allow the greenhouse to run on the same settings every day. Even if your greenhouse features a lot of automation, you have to react to current weather conditions, and use forecasts to look into the future and anticipate how to manage in response.

Other Actions

Depending on how much control you have over the conditions in your structure, there are a lot of other ways you can steer your crop. Think about how everything that happens to you over the course of your day affects your mood and the way you feel: whether you got enough sleep, had a cup of coffee, had breakfast, got stuck in traffic, had to skip lunch, and so forth. It's the same with plants: Almost every action you can control will affect how the crop is growing. Below is a list of things that are known to have a steering effect on a crop.

Note that if your plants are really far to one side of the continuum, you may need to use multiple techniques at the same time to steer them back into balance. Don't expect to see huge changes overnight. It takes at least a few days for the effects of steering to be noticeable. Once you have a crop balanced, the weather or other conditions may change, making it necessary to begin steering again. Like horticulture in general, crop steering is a constant balancing act, using the factors you can control to counter the effects of those you can't. See table 7.2 for a list of all the actions that can be used to steer the crop.

Temperature. As I've already discussed, the wider the day/night temperature differential is, the greater the generative effect is. This can be combined with a fast drop in day/night temperature to do pre-night treatments.

Humidity. Increasing the humidity to high levels, within the acceptable range for the crop, serves as a vegetative action. Lower humidity will make the crop more generative, because it has to work harder to transpire at an adequate rate, which is more stressful on the plant.

Watering. Any watering strategy that results in the roots being more uniformly moist over the course of the day will have a vegetative effect. For example, multiple small, evenly spaced irrigations will tend to have a vegetative effect. Fewer, larger irrigations with more drydown between waterings are more stressful for the plant and thus are also generative. So is waiting until later in the day to begin irrigation, and having the last irrigation of the day earlier, so there's a longer period between irrigations from one day to the next, with more drydown overnight.

Fertility. Soil-grown crops are more buffered to changes in the root zone, so adjusting soil fertility is not the quickest method of steering. However, excessive nitrogen levels will make plants too vegetative.

EC. Increasing the level of EC (salt concentration) in the root zone is moderately stressful for the crop and has a generative effect. Try incremental changes of plus or minus 0.5 mS/cm (millisiemens per centimeter, a measure of saltiness) from your normal EC to begin with and see if you notice an effect. Over the long term, running a higher EC may increase flavor and decrease yield (for a more detailed explanation, see the "Root Pressure and Watering" sidebar on page 59).

Carbon Dioxide. Higher levels of carbon dioxide have a more generative effect.

Pruning. There are a number of ways that pruning can affect the state of a plant. More leaf pruning will have a more stressful, generative effect on the crop, while keeping more leaves on will be more vegetative. Aggressive leaf pruning from the beginning can be one way to settle a crop that has started out too vegetative.

Harvesting. Leaving more fruit on the plant will have a generative effect, as a higher fruit load draws more strongly on the plant's assimilates. Conversely, if

A Generative Pruning Trick

A specialized generative pruning action can be used on tomatoes to divert more energy into the fruit, and since it relies on pruning, it works whether or not you have heat. It works on the principle that indeterminate tomato plants produce three leaves for every fruit cluster. Each leaf/cluster group can be thought of as a unit, with one leaf below, one leaf opposite, and one leaf above each fruit cluster. The technique is fairly simple: Remove the leaf opposite a flower cluster when the leaf is about 1 inch (2.5 cm) long. This redirects the energy that would have been devoted to growing that leaf to the flower cluster and the leaf both above and below. Some prefer to use their fingers for this; others like pruners. It doesn't matter as long as you can take the leaf off cleanly and not leave a stub.

Remember that leaves are energy sinks from when they first unfurl up at the top of the plant until they mature and become energy sources. So this technique rebalances the amount of energy; slightly more goes to the fruit than otherwise would have been devoted to leaves.

If you really want to geek out on the details, in a group of three immature leaves and the associated flower cluster at the head of the plant, the cluster gets about half the energy and each leaf gets about a third of the other half. Let's say the cluster gets 49 percent of the energy, and each leaf gets 17 percent (for a total of 51 percent for the leaves) of the total amount of energy the plant is going to invest in that section. When you remove the leaf opposite a developing flower cluster, a third of that leaf's energy is redistributed to each remaining leaf and the flower cluster, boosting the amount of energy the flower cluster receives closer to 55 percent, giving the

plants have a high fruit load and are growing slowly, harvesting more aggressively can liberate some energy to go back to the head of the plant.

Crop Registration

Crop registration is a way to track the progress of your crop that is particularly useful for helping determine if the crop is trending in a vegetative or generative direction. The idea is to pick some plants at the beginning of the season and measure their progress every week until the end of the season. Tie brightly colored flagging tape to the plant's string or use some other method that will help you remember which plants you're monitoring. It's also important to number the plants, because the week-to-week trends will be off if you don't keep track of an individual plant's statistics.

Make it a weekly task to measure the plant's growth and record the information. This is a great job for an apprentice or someone learning how to be a grower. The vital stats you want to keep track of are how much the plant grew in the past week, and the diameter of the stem 6 inches (15 cm) below the tip of the plant. Record how far from the tip the newest flower cluster is on each plant. You can also record how many flowers are on the cluster, and how many of them set. The first week you take a measurement, make a mark on the string at the tip of the plant; the next week, measure from that mark to determine how much the plant grew. Get a pair of calipers, measure 6 inches below the tip of the plant, and take a measurement of how wide the stem is at that point.

In addition to providing a record of crop growth, this data will show you how the crop is trending. You may be able to see changes in data before they are visible to the naked eye, so you can notice trends more quickly. If the rate of growth is slowing week after week, and the stems are getting thinner, it is a good indication the plants are becoming more generative. The opposite may indicate a vegetative shift.

generative part of the plant more than half of the total energy and gradually shifting the plant's energy in a generative direction. The exact percentage is less important than the fact that you're shifting the energy in favor of the generative parts of the plant.

You can start doing this technique as early as the first flower cluster if your plants are too vegetative from the beginning, or if you know from experience that they tend to be that way right out of the gate. To be an effective generative steering tool, this technique must be done when the leaf is very small; otherwise the plant will already have made its investment of assimilate in the leaf.

You can keep doing this on successive flower clusters until the plant settles down and becomes more balanced. If you are doing other generative steering on the plant, taking a leaf is the action you want to stop first as the plant comes back into balance. Taking a leaf out of the head will reduce yield over time, because you are reducing the number of leaves. On the other hand, being excessively vegetative has a greater potential to reduce yield, so it's definitely worth doing if other generative actions don't seem to be enough to balance the plant. When you remove an immature leaf, the plant will compensate by lengthening the leaves that remain to make up for some of the missing photosynthetic surface area.

It is not standard practice to take leaves out of a head of peppers or cucumbers. Since peppers are naturally more generative, they usually don't require this kind of measure to balance. And since the leaf-to-fruit ratio is 1:1 in cucumbers, the rebalancing effects of removing a leaf would be more drastic than in tomatoes.

Records from crop registration can be compared with weather and harvest data, as well as notes about what was going on in the greenhouse (pests and diseases, for instance), to help you decipher trends. You may be able to figure out whether a low yield trend is due to the crop being too vegetative, or to excessively high temperatures, or to both factors at the same time.

Different growers use different standards for what is considered vegetative or generative growth. For example, a 0.75-inch (1.87 cm) stem width at the flowering cluster might be considered vegetative, whereas 0.5 inch (1.25 cm) might be considered balanced. What's more important is the trend over time: Are stems getting fatter or thinner? Are plants starting to grow faster or slow down?

A good set of crop registration records will help you become more proactive by anticipating what will happen based on previous years' records. Having records and a historical perspective will help you predict when hot weather will hit, or when pests and diseases will show up, so you can be ready when things change. Make it someone's job to do crop registration every week and you'll have the data when you need it.

CHAPTER EIGHT

Grafting

Exactly how long humans have been grafting plants is unknown, but there are records of it dating back over two thousand years. Grafting is a natural process that takes advantage of the plant's ability to heal itself in order to combine it with another plant. You are probably familiar with grafted grapevines and fruit trees. Most if not all of the apples and grapes that you have ever eaten have been grafted. The same basic technique can be applied to annual vegetables. Grafting allows for the creation of a "double hybrid," taking the roots of one plant and the top of another, called the scion, and splicing them together to enjoy the best qualities of each united in one plant.

Grafting is the most important development in tomato growing since the commercialization of hybrids in the mid-twentieth century. This is a big claim, but grafting represents a second revolution in tomato hybridization, and has become a standard way for protected culture tomato growers to increase yields. The yield boost usually seen with grafting may come from increases in disease resistance, abiotic stress tolerance, and vigor. Because grafting can boost any variety's root disease resistance, it's a great way for organic growers – or anyone – trying to increase resilience without chemical usage.

At this point in time, grafting is not nearly as common in any of the other crops, though it can be done with eggplant, peppers, and cucumbers. Breeding work is ongoing to improve rootstocks for these other species. As better varieties are developed, grafting may come to have the same benefits, and become just as important, for fruiting crops beyond tomatoes.

Grafting is such an exciting technique because the improvements that come from using a good rootstock can be applied to almost any top variety, regardless of whether the top is hybrid or not. You can take your favorite old heirloom variety and give it a modern disease package and root system with grafting, without sacrificing fruit quality. You can also customize the top/rootstock combination for your own climate conditions. Let's say you really love Brandywine tomatoes, but your conditions tend to be very generative, and your Brandywines peter out before the end of the season. Grafting Brandywine to a strong rootstock has the potential to boost its vigor to keep it strong through difficult conditions.

The Benefits of Grafting

One of the biggest benefits of grafting is the yield boost. Your mileage may vary, but grafting most varieties will result in a higher yield than growing the top variety on its own roots. How much yield is increased depends on a lot of factors.

Grafting is a technique that will have more dramatic results under more adverse conditions. When I worked at Johnny's Selected Seeds, I did the grafting and ran rootstock trials at the research farm. For many of those years, the trials were conducted in an unheated hoophouse. I spliced a lot of different top varieties onto a lot of different rootstocks. Most combinations yielded a 30 to 50 percent increase over the ungrafted top variety grown as a comparison – and that was without

Figure 8.1. Grafting woody plants dates back thousands of years and is used for production of many crops, including apples and grapes. Applying the technique to vegetables is a newer development. Grafting occasionally happens in nature without any human intervention, as with these self-grafted crab apple trees. Public domain.

soilborne diseases taking a toll on the ungrafted plants. In other words, both grafted and ungrafted plants were healthy, so the difference in yield wasn't an effect of disease reducing the ungrafted yield. That experience has informed my basic advice to growers that grafting is worth it in most cases for the yield boost alone – even in the absence of disease pressure.

Still, it's important to understand that grafting can only improve the soilborne disease resistance of a plant – not the foliar disease resistance. It is a common misconception that whatever disease resistances the rootstock has will find their way into the top. But what you have to remember is that after a plant is grafted, each part of the plant – scion and rootstock – retains its own distinct disease package. Even though they are joined together, top and bottom still have their own disease resistances.

For example, a lot of people have asked me whether using a rootstock resistant to late blight will make the top variety resistant to late blight as well. The answer is an emphatic no. A scion grafted onto a late blight-resistant rootstock will be just as vulnerable as if it were grown on its own roots. This is why, when you look at the disease-resistance packages of varieties bred for rootstock use, they are mostly for soilborne diseases.

Along with overcoming stress from disease, rootstocks increase the resilience of grafted plants to abiotic stresses, meaning those that are not caused by pathogens, including flood, drought, cold, heat, fertility, and salinity stress. The increased level of vigor helps broadly with disease resistance, in that a more vigorous plant has a stronger constitution; it's able to grow faster and in some cases can outgrow disease. The higher vigor may also help keep plants robust in colder and hotter-than-ideal temperatures.

One reason why grafted plants are more resilient to abiotic stress is that they have larger root systems. This can help the plant deal with conditions that are too hot. A bigger root system has the ability to supply more water to the top of the plant when it is respiring quickly, helping to minimize the effects of heat stress.

The same principle may help a grafted plant deal with drought stress. Since a larger root system can draw water and nutrients from deeper than can a plant on its own roots, grafted plants may perform better than ungrafted ones in times of water and fertility stress. Thus you may be able to apply less fertility to grafted plants and get the same results.

Though tomato plants don't like to have flooded roots for any extended period of time, grafted plants may be more resilient to flooded conditions, possibly due to the higher level of resilience imparted by the rootstock. Where flooding is a common event, cross-species combinations to increase tolerance to wet roots are possible. Eggplant is more tolerant of having its roots flooded than are tomatoes. When flooding is expected in the tropics, growers often graft a tomato plant onto an eggplant to increase flooding resistance.

Though grafted plants may receive some added resistance to excess salinity through their tougher rootstocks, researchers are working on rootstocks that are highly salt-resistant. This would be hugely valuable in parts of the world where salinity is a common problem, such as the southwestern United States, the Middle East, and other arid regions.

Grafting is as much a plant breeding trick as it is a plant husbandry trick, freeing the breeder from having to cram all the ideal characteristics of a variety into a single plant. When you think of everything an ideal plant would have, it's a pretty tall order: good flavor, nice texture, attractive fruit, high yield, soilborne disease resistance, foliar disease resistance, strong vigor, manageable growth habit, long shelf life, et cetera. The list could go on. And in some instances, goals can be at odds with one another. For example, flavor and shelf life are frequently inversely correlated; many people find that the firmness bred into shipping tomatoes detracts from the flavor.

Grafting allows breeders to focus exclusively on belowground traits in the rootstock variety, and on aboveground traits in the top variety. So breeding for the scion emphasizes fruit quality, plant habit, foliar disease resistance, and easy maintenance, while breeding for the rootstock emphasizes root mass, soilborne disease resistance, and level of vigor. Breeders can forget about fruit quality altogether in rootstocks. In fact, if you let a rootstock variety grow without grafting it, the fruit produced is usually late, low yielding, and borderline inedible. All the energy in rootstock breeding goes into the belowground parts of the plant.

This gives a lot of flexibility not only to the breeder, but to the grower as well. It means you can make your own custom combinations of top and bottom varieties suited to your growing conditions. Is your favorite heirloom tomato a little weak in the vigor department? Graft it onto a super-vigorous rootstock. Do you want to grow the same variety in the field and in protected culture but you face different soilborne disease problems in each location? Use a different rootstock with the appropriate resistances in each place.

Grafting is also a specialized plant breeding trick in that it maximizes the potential of hybrid vigor. Hybrid vigor is the tendency of hybrid plants to be more vigorous than their open-pollinated counterparts. The more distantly related the parents of a hybrid are, the more pronounced the effects of hybrid vigor are. The analogy in the animal world is the mule, which is the result of a donkey bred to a horse. This type of cross of the same genus/different species is called an interspecific cross. The resulting mule is stronger than either of its parents.

In vegetable grafting, most of the rootstocks used for solanaceous and cucurbitaceous crops are interspecific crosses in order to take advantage of hybrid vigor. Interspecific crosses are not the same thing as genetic engineering. Just as mules are produced in the wild if the right horse and donkey cross paths, interspecific solanaceous or cucurbitaceous crosses could and no doubt do happen on their own. In plant breeding these crosses are made by hand, with traditional plant breeding methods involving the physical transfer of pollen.

In the case of tomatoes, interspecific crosses don't usually happen on their own, simply because tomatoes tend to be inbreeders, with self-pollinating flowers that are usually pollinated by the time the flower is open. So these crosses just need a plant breeder to play matchmaker and come up with the right combination for the next great rootstock.

As grafting catches on with more growers, more energy is going into breeding rootstocks, with new ones becoming available at an increasing rate. The trend in rootstock breeding is to develop more customized combinations: say, rootstocks with disease packages suited to a specific region, to deal with a specific stress like salinity, or rootstocks specifically for small- or large-fruited varieties. I recommend growers give grafting a try for tomatoes or any other crops with rootstock varieties that promise a yield boost, or that may help solve production problems.

I occasionally hear reports from growers who say they've grafted plants without any yield benefit. But given my experience, in which grafted plants outyielded their ungrafted checks in every single trial I ran over a number of years, this is hard for me to reconcile. My guess is that some of these growers had grafts that did not heal properly. It's possible for a plant to survive a bad graft even though its vascular structure has not properly healed, and in these cases the plant will not reach its full potential. These plants are easily eliminated from a flat of grafted plants because they do not grow as quickly as the properly healed plants.

Grafted plants also may not live up to their full potential if they become extremely vegetative at the expense of fruit production. I have had plants go excessively vegetative on me, which can hurt long-term yield. If this happens, use the crop steering techniques described in this book to balance the plants.

The last reason why rootstocks may not live up to their potential is if conditions are already very close to optimal. The worse conditions are, the bigger the difference rootstocks can make. If conditions are already near ideal, grafting will have less of an effect. In some of the greenhouses I have visited where growing conditions are optimized, the yield boost from grafting is closer to 10 percent. Ultimately, the promise of grafting is so great that it's worth trying to find out whether or not it suits your own particular growing style.

Choosing Rootstocks

One of the most basic considerations when choosing a rootstock is whether or not you have soilborne diseases. If so, it's worth finding out whether rootstocks are available that are effective against those diseases. Rootstocks are now available with resistances to many of the common soilborne diseases.

The other main consideration is vigor. The level of vigor in the rootstock variety will greatly influence the vigor of the grafted plant. Most commercial rootstocks are bred to boost the level of vigor in the scion. They range from extremely vigorous rootstocks that elevate the level of vigor in the finished plant to a great degree, to moderately vigorous rootstocks that will boost the plant somewhat but not make a huge change. The vigor of the grafted plant is a function of the vigor of the top variety plus the vigor of the bottom variety. For example, a very vigorous top variety grafted onto a very vigorous bottom variety might result in an overly vigorous combination. For this reason, it's a good idea to test new combinations before planting a whole lot of a new scion/rootstock combination.

I've heard of some growers using non-rootstock varieties as rootstocks in an effort to save money. By picking a cheap variety with resistance to their soilborne diseases, they figure they're getting the disease resistances without paying the premium for real rootstock seed. While this seems like it makes financial sense, it is penny wise and pound foolish, and a lot of trouble to go to without getting the full benefits of grafting. In an attempt to save a few cents per seed on rootstock, these growers miss out on the increased vigor, yield boost, and stress tolerance of using a real rootstock variety. I know; I've tried using vigorous fruiting varieties as rootstocks. Even the most vigorous fruiting variety is less vigorous than a real rootstock.

Here's a simple calculation you can do to figure out whether it's worth it to buy the rootstock seed or not: Take the cost of the real rootstock variety you would use and subtract from this the cost of the non-rootstock variety you're thinking of using. Then think about the yield benefit you might realize from using a real rootstock variety. Now calculate how much more money you would be making per plant if you got the extra yield benefit of a vigorous rootstock. If the amount your income would increase from using a real rootstock is greater than the cost of the rootstock seed, then it's worth it to shell out for the real one. In most cases, the benefit of using a rootstock variety is much greater than the cost.

Grafting Process Overview

Grafting is both an art and a science. It's not rocket science, but it's a process that has to be managed carefully for consistent success. In other words, it can be a pain. To some growers (like myself), the process of cutting their plants apart and splicing them back together has a certain nerdy appeal, in addition to the benefits that it can confer. For other growers, cutting up seedlings is the last thing they would want to do.

Over the years, I have talked with many growers who taught themselves how to graft. I've also met many growers who wanted the benefits of grafted plants but weren't able do the grafting themselves. Maybe they couldn't get the hang of the process, or didn't have the proper facilities, or just didn't like doing it. Luckily for these growers, grafted plants are becoming commercially available on a smaller scale, which means they can try grafted plants from someone else before they invest in learning the procedure.

There are benefits for growers who want to do their own grafting, including a wider choice of scion, rootstock, and propagation medium, and freedom from minimum order sizes. Some growers simply enjoy having control over all stages of their plants. Here's an overview of the grafting process, so you can decide whether it's something you want to do yourself or not.

Regardless of what method you use to graft or what species you are working on, all grafting involves a few distinct steps:

First, you have to propagate seedlings.

Then, you have to splice the seedlings together such that both varieties will heal together to form one plant.

Next, you have to place them in a healing chamber to ensure the plants survive the process.

Finally, you have to gradually reacclimate the grafted plants to normal greenhouse conditions.

The Healing Chamber: The Most Important Step

The reason I'm bringing up the healing chamber before getting into the actual grafting is to emphasize that you need to have your tools and healing chamber ready *before* your plants are ready to graft. Preparing ideal healing conditions will make all the difference in the success or failure of your grafts. A good healing chamber can make up for average grafting skills. The most perfect grafts in the world, however, will die in a bad healing chamber. Grafting is like performing surgery on your plants, and the healing chamber is the intensive care unit. It doesn't matter how good a surgeon you are; if there's nowhere for your plants to recover, they're not going to make it.

The basic requirement of a healing chamber is the ability to control the levels of light, temperature, and humidity as precisely as possible. The more closely you can keep the conditions within the ideal range, the better your chance of success. Depending on the size of your operation, a healing chamber can be the size of a dome over an individual flat, a closet, a room, or an entire greenhouse. I've seen a lot of different setups used successfully over the years. Healing chambers don't have to be fancy. Anything you can put together that keeps those three factors at the ideal levels can work.

I prefer healing grafted plants indoors, not in greenhouses, because there is more climate variability in a greenhouse. The grafting process will go more smoothly if you can hold the ideal temperature for each step steady, regardless of what's going on with the weather outside. Seedlings that have just been grafted have very little resilience; after their vascular structure has been severed and before it is healed, grafted seedlings have very little ability to adapt to fluctuations in the environment. The other disadvantage of grafting in greenhouses is that they tend to be breezy places. Wind is one of the worst things for newly grafted seedlings. Plants that do not have a good vascular connection cannot replace the moisture in the leaves that is lost to wind.

It is possible to graft in a greenhouse, however, as long as you manage the complicating factors. If you have to graft or heal in a greenhouse, turn off fans and eliminate as many sources of wind as possible. Try to graft on an overcast day, or put shade cloth over the structure to limit the heating effects of the sun. Healing chambers can be constructed within a greenhouse by making mini tunnels on bench tops or on the floor, where light, temperature, and humidity levels can be controlled.

One of the most common ways to make a healing chamber is to put a humidity dome over individual flats, and put the flats in an environment (a closet, room, shipping container, or the like) where light and temperature can be kept in the optimal range. You can also construct a healing chamber by making a tent to maintain the humidity over one or more flats, and then control the light and temperature in the tent. The tent could be as simple as a piece of plastic over a couple of flats, or as extensive as a tunnel covering an entire bench.

To make a tabletop bench, hoops for floating row cover or other heavy wire can be attached to the top of a wooden bench by drilling holes for the wire in the top of the table, and inserting the ends of the hoops into the holes. Or hoops can be attached to the side so that they're suspended over a bench much as row cover hoops straddle a bed in the field.

Temperature

For whatever species you are grafting, there is an ideal temperature for healing, so having some kind of thermostatically controlled heat will make it easier to maintain the desired temperature. This can take the form of a heated room, a space heater, or even bottom heat to maintain the temperature.

Humidity

A high level of humidity is necessary after grafting to keep the plants from wilting while they are healing. Humidity can be achieved by two methods. The first is to create humidity in an enclosed area by misting and then seal it to maintain the humidity. The most common example of this is misting an individual humidity dome over a tray of grafted plants. The second is to constantly replenish the humidity in an enclosed area. The easiest way to create and maintain a constant supply of humidity is to use a cool mist humidifier, as in a tent or plastic tunnel.

Light

After the grafting process, seedlings need to go into a reduced- or no-light environment. Depending on the desired light levels, this can involve putting shade cloth or tarps over tents or tunnels in a greenhouse. Yet another advantage of healing indoors is that if the plants need to be in the dark, you can just turn the lights off. The point of being in the dark is to keep the plants

Figure 8.2. A tray of healed seedlings that have just emerged from healing, with lights close overhead to keep them stocky.

from doing anything as they start healing, including photosynthesizing. Remember, the healing chamber is the plant intensive care unit. They need to take it as easy as possible.

Propagation Mediums for Grafting

Regardless of which grafting method you use, it's frequently useful to be able to pull the plants out of their cells to sort them. An example of this would be sorting the biggest plants in flats for grafting first, and the smaller plants for grafting later. Sometimes this needs to be done before the roots have filled the cell enough to hold a soil plug together. When you're considering which planting medium to use for grafting, therefore, note that it's advantageous to use a medium that can be handled easily without falling apart.

This will also help with cleanliness. If dirt or pathogens get in the cut between the two plants, this will prevent them from healing together. With very loose mediums, you run the risk of your root ball falling apart and getting soil all over your hands and the plants. This may not sound like a big problem, but it will slow you down, and it can compromise the quality of the seedlings if the root mass falls apart.

I've grafted with seedlings in loose soil, and it can be very difficult and time consuming to keep it off the stems, your fingers, the razor blades, and so forth, if the root balls are falling apart. Hydroponic growers do not have this problem, because they have always used solid mediums to start their plants. But there's an interesting option that has recently come on the market called the Ellepot. It's a stable plug made of soil, held together with a paper wrapper. It's customizable because different potting soils

Figure 8.3. Grafted plants that are done healing. This batch got leggy due to condensation on the inside of the humidity dome blocking some of the light.

can be used to fill the plug. It's a way to both use soil and be able to handle the plug without a lot of mess or falling apart. At the time of this writing Ellepots are not acceptable in organic systems, but that could change in the future. You can also use a soil blocker to make blocks out of potting soil. Just make sure that your potting soil is a good mix for soil blocking; mixes that are not sticky enough can fall apart when watered or handled.

Organic growers who need a solid plug have a number of options. Rockwool is out because it is synthetic, but cocoa coir plugs are a good choice; just make sure that your certifier accepts the brand you wish to use. Some certifiers view Oasis cubes as an inert medium that doesn't contribute anything to the plant, and thus acceptable for organic production. But they are not OMRI-listed so as always, check with your certifier to make sure they're acceptable for organic production. Peat-based foam plugs may also be available that can be used in organic systems. All of these options can be potted onto a larger soil block for growing to transplant size after they have been grafted.

Propagation Planning

Plan on adding two weeks to your normal propagation time frame for producing a grafted transplant. Growth comes to a halt when a plant is grafted and gradually resumes during the healing process. For example, if you usually plan on planting an eight-week-old ungrafted tomato seedling, plant a ten-week-old grafted seedling.

It's very important to overseed when grafting, so you have more plants than usual. Most propagators I know overseed by anywhere from 50 to 100 percent. This is necessary to account for the possibility of poor germination, some percentage of plants that will not match because of varying stem size, and some plants that will not survive the grafting process. Given the expense of greenhouse varieties, you can also try overseeding a little more by the cheaper variety and less by the more expensive variety. For example, if you are using a really expensive scion, you could overseed the scion by 25 percent, and overseed the rootstock by 50 to 75 percent. On the other hand, if you are grafting with a top variety that's relatively cheap, you might want to overseed the cheap variety by a lot (like 200 percent) and overseed the rootstock by only 25 percent. This may seem like a lot of overplanting, but there's nothing more frustrating than not being able to do enough grafts because you've run out of matching scions and rootstocks.

Based on my experience working with heirlooms I have observed that the seeds tend to come up over a longer germination window than most rootstocks. I recommend planting two to four times as many seeds for heirlooms as the final number of plants you need. This may sound excessive, but having the right number of seedlings when it comes time to transplant is priceless. And heirloom seed is usually really cheap.

Materials

There are some materials commonly used with all crops and grafting methods.

Cutting Tools

Old-fashioned double-sided razor blades – the kind with the cutout in the center for the razor – are the best for cutting seedlings. They are much thinner and sharper than the single-sided razor blades, and with a slight sawing motion they can cut through stems with almost no resistance.

The best way to use these blades is to snap them in half so you have a thin, single-edged blade, which is more maneuverable when working on seedlings than an intact blade. You have to be extremely careful when working with these blades, because they are so sharp. Get the kind that are sold in a paper cover, because you can fold the blade inside its cover until it snaps in half, and then use the single-edged blade like a miniature knife.

An alternative way to make the cut for top grafting is to use a miter-cut grafting knife. This tool may make the cuts more uniform for those whose freehand cutting skills aren't so good. It can also be useful in speeding up the process if you have a lot of grafts to make. It works like an anvil pruner, with the stem cut against the anvil opposing the blade. There is a groove in the anvil for lining up the stem, and the blade is set so it cuts the stem at the proper angle when it's in the groove.

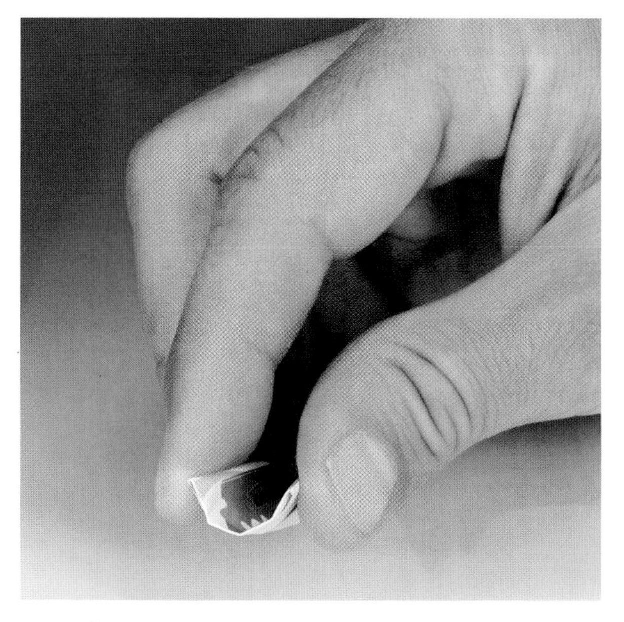

Figure 8.4. This is how you snap an old-fashioned razor blade in half to use for grafting. Bend it like this until it snaps in half. It's very important to leave it inside the paper cover or you could cut yourself. Photo courtesy of Azad Photo.

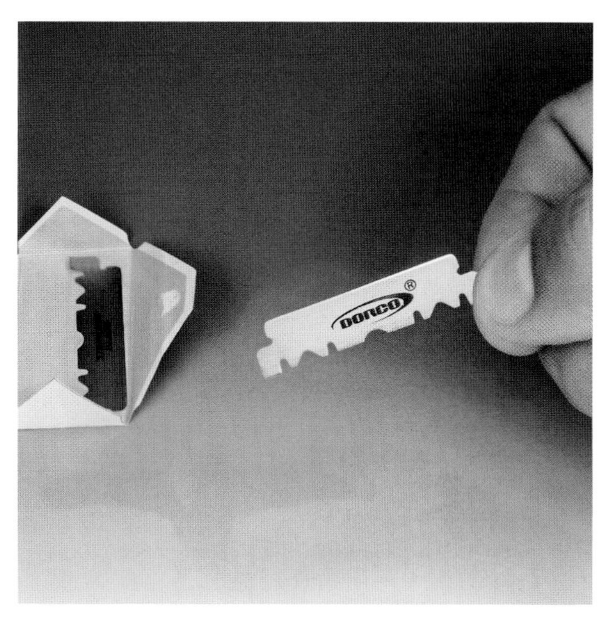

Figure 8.5. The finished product: a very sharp and thin blade that is more maneuverable than the double-edged razor blade. Photo courtesy of Azad Photo.

Figure 8.6. This is how the miter-cut grafting knife works. Squeeze the arms together and the blade severs the top against the anvil at the same angle every time. Photo courtesy of Azad Photo.

Grafting Clips

Top grafting clips are made of silicone, and are basically tubes with a slit on one side and ridges on the other to make them easier to hold. They come in a variety of diameters to accommodate different stem sizes. Some are color-coded by size, which can help when you are working with several sizes. Instead of ridges, some grafting clips have a loop of silicone opposing the slit, which can be used to insert a stake to keep the plants from keeling over post-graft.

Since grafted tomato plants cannot be planted deeply, they are prone to flopping over as they grow. Put a stake in the silicone grafting clip if it has a loop, or rubber-band the plant to a stake to keep it from flopping. Some places that sell grafting clips also sell stakes, or you can use coffee stirrers or bamboo shish kebab stakes.

Spring-loaded grafting clips are made of hard plastic with a small metal spring. They come in various sizes for different stem diameters. They work like miniature clothespins, gently clamping two cut stems together while they heal. These are good for approach grafting and cucurbit grafting.

Figure 8.7. Domes are used to maintain humidity on a light cart in a healing chamber.

Humidifiers

If you are using a tent to heal multiple flats at a time, you will need some type of humidifier. For a small- or medium-sized tent, a cool mist humidifier for home use should work. For humidifying individual domes, a pump sprayer is preferable to a squeeze bottle sprayer. The continuous spray generated by a pump is gentler on freshly grafted plants and less likely to knock the tops off.

Solanaceous Grafting

There are three main methods for grafting the solanaceous crops of tomatoes, eggplant, and peppers: top grafting, cleft grafting, and side/approach grafting. Each method has its own advantages and disadvantages.

Top Grafting

Top grafting is the fastest method for solanaceous crops and the least consumptive of time, materials, and space. The downside is that it's also the least forgiving. Because the plant's top and roots are completely severed, they are likely to die if the healing process doesn't go smoothly.

This method has a steep learning curve. You will kill some plants in the beginning. For this reason my best advice about top grafting is to learn how to do it on some expendable plants. Become confident in your skills before you try to graft the plants you need for production. Start practicing long before you need to plant. That way if you don't get the hang of it the first time, you can try again. Get out some old leftover seed, or buy the cheapest seed you can find just to practice on. I don't want to make this sound intimidating, because most people can learn to graft. Just don't learn how to do it on plants you need to plant.

Timing

Two aspects of timing are important for grafting. First, as mentioned above, you will need to plant seeds about two weeks earlier than usual. For example, if you normally plant an eight-week-old transplant, start your seeds for grafted transplants ten weeks before you plan on transplanting.

The other aspect of timing for grafting success is when to start each batch of seeds, rootstock and scion, in relation to each other. This is very important since one of the biggest factors determining the success rate of top grafting is how well matched the stems of the rootstock and scion variety are. They don't have to match exactly, but the closer they are to the same size, the greater the success rate will be.

Timing the plantings of rootstock and scion so they are both the same size when it comes time to graft can be tricky. In a perfect world, you would be able to plant the rootstock and scion on the same day and have them match up two to three weeks later when it comes time to graft. In reality, though, varieties often take different amounts of time to reach the same size. The major factors to take into consideration when timing plantings are time to germination and growth rate.

In my experience, a lot of the greenhouse tomato varieties germinate and grow at a similar rate to many of the rootstocks. So some varieties can definitely be planted on the same day. The best way to measure the variability in growth rates is to do a small planting of each variety you want to use a few weeks before you plan on grafting, and note how quickly they reach grafting size. An easy way to tell when they have sized up without actually cutting them is to take a grafting clip of the size you plan to use, and put it on an intact stem. When the clip is snug, it has reached the desired size.

Keep track of the growth of the scions in relation to the rootstock you want to use. For example, a scion that reaches grafting size two days before the rootstock could be planted two days after the rootstock, and vice versa for a slow-growing scion. This may seem like a pain, and it is, but it will save you a lot of grief and improve your success rate if you take the time to make sure your stem sizes match when you go to graft. Cherry and grape tomatoes tend to have thinner stems, so they may need a few more days to size up than most rootstocks.

Planting

Again, it is necessary to add two weeks to your propagation time line when planning on using top-grafted transplants. So adjusting your planting dates backward accordingly is the first consideration when it comes to starting seeds for grafting. Depending on how big you want the seedlings to be when it comes time to graft, anywhere from a 50- to a 128-cell tray is a fairly standard size for top grafting. Use whatever size you like to propagate tomatoes in.

The most common stem sizes for top grafting are between 1.5 and 3 millimeters. Most of the silicon clips made for top grafting fall in this size range, but it is possible to do it at smaller and larger sizes. I find plants much smaller than 1.2 mm difficult to work with. And the stems of plants in the upper size range may start to get woody, which makes them more difficult to cut accurately.

The other reason I like a plant around 1.5 mm is that at this size the plants don't have a lot of leaf area yet. This makes them easier to handle and reduces the amount of transpiration, which means they will take longer to wilt after being cut.

My advice is that when you learn to graft, try stem sizes in the range of 1.5 to 2 mm. Since I prefer grafting at 1.5 mm, I like to have a bunch of 1.5 mm clips on hand when I start the process. But I always make sure to have a bag of 2 mm clips as well. That way, even though I plan on grafting at 1.5 mm, I have some bigger clips on hand for when some plants get away from me and reach the next largest size. If I didn't have them on hand and had to order the next largest size by mail, the plants would likely get too big while I waited for delivery.

When to Top Graft

Start grafting as soon as the largest plants reach the desired size. Seedlings of this size are growing at a very fast rate; I am always amazed at how quickly they go from the right size to too big.

When you have plants that are the right size to top graft, it's wise to give the plants a good watering the day before, so they can dry out a bit before grafting. You want the medium to be moist but not sopping wet. It is possible for rootstocks that are very wet to have too much moisture flowing to heal properly. This is unlikely, but it can happen. The other reason to water the day before grafting is because freshly watered plants will have wet leaves and may be flopped over from the water. So if you start grafting, then realize that your medium is too dry and you need to water during grafting, water from the bottom by placing your flats in water. It is much easier to select plants with matching stems when they are not bent over.

If your plants aren't very uniform in size, you may want to sort them before grafting. For example, you may have a tray of seedlings where one-third of them are the right size and the rest are too small. If you put the ones that are ready to graft in a tray by themselves, you will speed up your grafting since you don't have to sort plants and graft at the same time. This is where having plants in a soil block, solid plug, or cube that won't fall apart during sorting comes in handy. Being able to graft more quickly because your plants are pre-sorted may increase your success rate, since the faster you can graft the faster you can get them in the healing chamber. The longer plants sit around after grafting before going into the healing chamber, the greater the likelihood that they will wilt.

Take seedlings out of the greenhouse a few hours before grafting, and put them in a cooler, less sunny environment to slow their respiration down. If you take plants straight from a hot, breezy greenhouse and start cutting them immediately, they will still be respiring fast. The faster they're respiring, the faster they will wilt after being cut.

Graft the plants and get them into the healing chamber without ever letting them wilt. Again, remember that grafting is the equivalent of doing surgery on your plants, and the healing chamber is the intensive care unit. Recently grafted seedlings have very little resilience because their vascular system has been severed. If they wilt before getting into the healing chamber, they may never be able to recover. Only graft as many plants at once as you can get into the healing chamber before wilting. For example, if you are grafting and you notice your first graft wilting by the time you are doing your twenty-fifth plant, you may want to start putting your plants in the healing chamber in groups of twenty.

If seedlings have a lot of foliage when you go to graft, you may want to remove some of the leaves first. You can cut the whole leaf off at the stem with a razor blade. This will make the stems more accessible and easier to handle, and less leaf area will reduce the rate of respiration. One of the advantages of using smaller plants is that they don't usually need to have leaves removed. Leaf removal can also be used as a last-ditch effort to save plants that are wilting. If you have a batch of newly grafted plants that are wilting, you can go so far as to remove all the leaves as long as you leave the growing point. Still, it's much better to remove leaves before grafting if you suspect they're too leafy.

If possible, do your grafting out of direct sunlight, because the light and heat will cause the plants to respire more quickly. A room temperature somewhere in the 70 to 75°F (21–24°C) range will be comfortable for both you and the plants. Exact temperature while grafting is not important, but the plants will respire more slowly if you aren't grafting in a hot, breezy area. If you are grafting in a small, enclosed area, you can use a humidifier to make the plants less prone to wilting during the process.

Making the Cut

Before you start cutting, it's important to have a clean work area. It needs to be free of dirt and pathogens, which can easily get into the plant at the grafting wound. And it's important to have your domes or healing chamber ready to receive the plants. If you're using a humidifier in a chamber, you will want to have that going before you start grafting, so the humidity is already high when your first batch of grafted plants goes in the chamber.

Figure 8.8. This is a good-sized seedling to use for top grafting. Photo courtesy of Azad Photo.

Figure 8.9. Here is the top being cut for a top graft. Sever the top below the cotyledons freehand using a razor blade snapped in half, or a grafting cutter tool like this. Photo courtesy of Azad Photo.

Figure 8.10. This is what the rootstock should look like with the top severed. Photo courtesy of Azad Photo.

Figure 8.11. Cut the scion to match the rootstock in stem diameter and the angle of the cut. You can compare the cuts to see how well they match like this. Photo courtesy of Azad Photo.

Figure 8.12. Put the scion into a silicone grafting clip. Photo courtesy of Azad Photo.

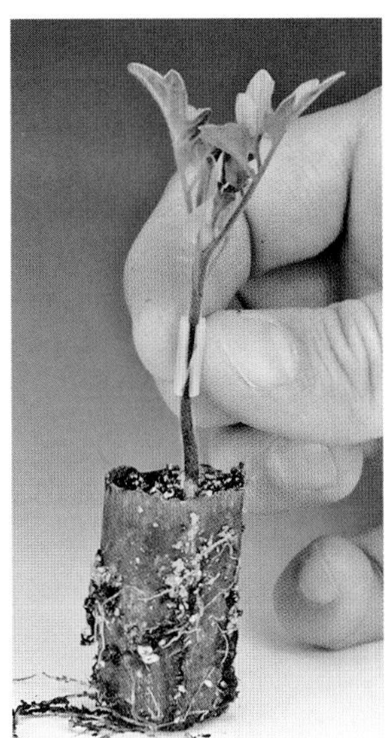

Figure 8.13. Lower the scion onto the rootstock. You should be able to see the cut slanting diagonally across the slit in the grafting clip. Photo courtesy of Azad Photo.

Figure 8.14. Make sure the scion is firmly in contact with the rootstock. Photo courtesy of Azad Photo.

It doesn't matter whether you cut the rootstock or the scion first, as long as you remember to keep track of which is which. This can be difficult because they may look very similar. I use the mnemonic device of always keeping the rootstock on the right, because if I ever get confused and wonder which is which, I think *R* for *Rootstock* on the *Right*.

Pick out two plants with matching stem diameters. If you're using a razor, hold the blade between your thumb and forefinger. If you find this difficult to grasp you could put the blade in a holder. Pick a plant and make a cut at a steep angle, 60 to 70 degrees across the stem of the plant, just below the cotyledons, the first leaves to emerge upon germination. You want the angle of the cut to be greater than 45 degrees because a steeper angle provides more surface area for the plants to heal together.

Completely sever the top from the bottom. A gentle sawing motion will help the blade glide through the stem and not bind up. As soon as you've cut a plant, throw away the part that's not needed so there's no way to mix them up and graft the wrong parts. Immediately toss the cut tops from rootstocks, and the roots from scions.

Next take the other plant, and cut it in the same manner, making a cut that matches the other plant as precisely as possible. If you notice a drop of moisture coming out of the stem after your plants are cut, this is a good sign. There needs to be a flow of moisture up from the rootstock into the scion in order for the graft to heal.

If for some reason you notice that your plants are very dry once you start grafting, go ahead and water them — plants that are in drought stress will not graft well. Use bottom water, by dipping flats in standing water or pouring water into the tray if it holds water, to avoid flopping the plants over.

Cut Placement

Try to make the cut at least 1 inch (2.5 cm) above the medium. The closer the graft union is to the medium, the greater the chance that the top variety will form adventitious roots and grow into it. Unless your plants are really leggy, make your cut as high up on the stem of the rootstock as possible, just below the cotyledons. If your plants are leggy, make the cut 1 to 1.5 inches (2.5–3.75 cm) above the surface of the medium.

On the scion, you want to make the cut farther below the cotyledons than on the rootstock. Since you will be keeping the cotyledons on the scion, you need enough stem below them to accommodate the grafting clip. One nice thing about grafting is that if your plants are leggy, you can resize them below the cotyledon by making the cut wherever you would like it. On the other hand, if your plants are very short, there's nothing you can do to increase the amount of space between the medium and the cotyledons. Short plants will have to be handled carefully to keep the graft union from becoming buried.

Splicing

Once you have two plants with the same stem diameter cut at the same angle, the next task is to join them. If you are new to grafting or have any doubts about how well the two plants will match, you can touch the cut on the scion to the cut on the rootstock without a clip to check their match. The best grafts fit together so well they look like they were never cut.

If you notice that the diameter or angle isn't perfect, you have a few options. If they are less than perfect but pretty good, you may decide to go ahead and put them together. Splices do not have to be perfect, but the better they are, the better your chance of success. How far from perfect the splice can be and still survive depends on your healing conditions, and is something you have to learn for yourself from experience.

If the diameter or angle is really poorly matched, it may be worth trying again. It is possible to recut one of the plants if the angles don't match. If the stem diameters are really different, you just have to find another plant that's a better match. Depending on the conditions in your grafting area, you probably have a few minutes to find another match before the cut surfaces dry out.

The silicone clips many grafters now use for top grafting were developed from surgical tubing that was originally used to join the stems when this method was being developed. In fact, you can make your own grafting clips out of surgical or other sterile tubing with the right interior diameter. It does help if the material you use is clear, so you can look through the tube and verify the position of the cut surfaces. To make your own grafting clips, make a slit down one side of a length of tubing. Then cut the tube into lengths of approximately 0.5 inch (1.25 cm). The clips made specifically for grafting are easier to handle because they have tabs opposite the slit to help hold and open the clip. But otherwise homemade clips can work just as well.

Once you have decided to go with a particular top/bottom combination, it's time to put a grafting clip on. I prefer to put the scion in the clip first and then lower it onto the rootstock, because I think they're easier to handle that way. But it doesn't really matter which plant you put the clip on first. To hold the plants together properly, the clip should fit snugly on the stem. If the stem is slightly larger than the hole in the clip, it may help to open the clip up by squeezing the tabs opposite the slit. Since the cut was made at an angle, you want to align the stem in the clip so that you can see the slant of the cut crossing the opening in the clip (see figure 8.13). If you put it the other way, with the point of the cut at the slit in the grafting clip, the flap of tissue at the point tends to dry out. It probably won't cause graft failure, but it's easy enough to prevent.

One little thing I like to do to make sure that the cut surfaces are in good contact with each other is to push the scion down in the grafting clip a little more than halfway. Then when I put the grafting clip with the scion in it on the rootstock, I apply some downward pressure on the clip until I see the scion being pushed upward by the rootstock from below. This lets me know the cut surfaces are in good contact with each other. This is not a necessary step, but it's one I like to do to verify that the cut surfaces are firmly touching. I try to end each graft with the cut surfaces meeting in the middle of the clip, so there is an equal amount of clip holding each plant.

Handling Newly Grafted Plants

Once you have the first plant grafted, you want to move on to the next one quickly so you can get them in the healing chamber as quickly as possible. You have a limited period of time before newly grafted plants start to wilt. You need to get them in the healing chamber to stop the plant's transpiration before this happens.

If while doing a batch of grafting you notice that the first plant you grafted is wilting, get the whole group into the healing chamber as quickly as possible. If the first one is wilting, the others probably aren't far behind. If grafted plants wilt a little bit they may recover, but it is ideal for them not to wilt at all. Newly grafted plants have very little resilience until they have healed.

It's also important not to bump grafted plants until they have healed. If the cut surfaces get jostled so they are not in contact with each other, the plants will not heal. It can help to space plants out in the tray going into healing, rather than using every cell. If you are doing a lot of grafting and need to use every cell, another solution is to use trays that have removable strips of cells. Instead of spacing plants out over the entire tray, you can take each individual strip out and deal with one strip at a time without worrying about bumping adjacent plants. You can buy strip trays, or make your own by piling up regular flats and using a circular saw to cut each tray into strips.

Depending on how fast a grafter you are, and how dry your air is, you need to move plants into the healing chamber in batches that are as large as possible to minimize the number of trips, but not so large that the first plants are wilting by the time you are grafting the last ones. This is another way using strip trays can come in handy. They can allow you to have a smaller batch size, and transfer a single strip of plants into a dome or healing chamber without having to deal with an entire flat at a time.

Speeding Up the Grafting Process

Once you have mastered the basic top graft, there are some methods you can use to speed things up. One method involves cutting a bunch of tops at once, and putting the stems in water to keep them moist until you can graft them. Then you can cut a bunch of rootstocks and do all the splicing at once.

This speeds up the process because you are not switching between tools and jobs on every single plant, going back and forth between cutting and grafting. Hygiene is

Figure 8.15. Trays of freshly top-grafted tomatoes that are healing.

especially important with this method, however, since if there is a pathogen on any of the plants, and they go in the water with other plants, they will all have it by the time you're done grafting.

Another thing you can do to speed the process up is to pre-sort your plants. If you're working on flats of seedlings of mixed sizes due to variations in germination and growth rate, you will waste a lot of time hunting around for seedlings of matching sizes. If you pre-sort the seedlings by size the day before grafting, this will save you a lot of time.

Cutting the rootstock and scion at the same time can also speed things up. In addition to the materials for top grafting, you also need a clean cutting surface, because you'll be cutting the plants on their sides. Something that the razor blade won't cut, and that is easily sanitizable, like a flat piece of metal or glass, works well. For this method, pick a matching scion and rootstock and lay them with their stems side by side over the cutting surface, root balls facing in opposite directions. Hold the stems together with one hand, and cut through them both at the same time with your other hand. Because you are cutting both stems at the same time, the cuts should match. After making the cut, you can sweep half of the plant material into the compost. For example, if you had the root ball of your scion on the left and your rootstock on the right, after you make the cut you will be left with the roots of your rootstock and the top of your scion on the right. You can quickly sweep all the plant material on the left into the compost. This method gets really fast when you set up multiple grafts at once. You can set up ten scion/rootstock combinations, make ten quick cuts, and splice all ten back together quickly.

Healing

The conditions you want to maintain for the first twenty-four to forty-eight hours after grafting are 95-percent-plus (as high as possible) humidity, complete darkness, and 80 to 82°F (26–28°C). You want it so warm because that is the temperature at which scar tissue forms the fastest for solanaceous crops. You want it dark because you don't want the plants to do anything, including photosynthesizing. Just as doctors advise patients recovering from trauma to rest, you want your healing plants to do as little as possible.

As far as humidity goes, you don't need to have a humidity meter for this. Just make it as humid as possible. When the air is at 100 percent humidity, transpiration stops. This is what you want for healing: You want the moisture to stay in the leaves until they can get it from the roots again.

If you are using domes to keep the humidity high around individual flats, use a pump-type spray bottle to mist the interior of the dome and mist the plants before putting the dome on. I like to mist the interior of the dome, mist the plants, and then shoot a little more moisture into the dome as I'm hinging it down over the plants. Be careful not to knock the tops off the plants with the sprayer.

If you're using a larger structure like a tent to keep the humidity up, run a cool mist humidifier constantly inside the chamber. The chamber doesn't have to be perfectly sealed since you have a continuous source of humidity, but you want it to be sealed pretty tightly, otherwise you will get areas of lower humidity near any holes in the chamber.

It's a good idea to check on the plants in the healing chamber during the first twenty-four to forty-eight hours even though they shouldn't need much. Make sure there are no leaks reducing the level of humidity. This is particularly important if you're using humidity domes where the humidity is not being replenished.

Sometimes if there isn't a good seal between dome and tray, or any other gap where humidity is escaping, it will be apparent because the rest of the dome will be foggy, but it will be clear near the leak where humidity is lower. Keep an eye out for escaping humidity because it's important to replenish this if you're using domes. If you find a leak, take the dome off and rehumidify, and see if you can fix the leak. If the humidity goes down too quickly after grafting, the plants may die.

After the first twenty-four to forty-eight hours, keep the humidity high but reintroduce the plants to light. Use light that is not as strong as daylight at first. When I am using humidity domes, I put the trays on a light cart that has four fluorescent tubes over the plants, and remove two of the tubes so the light is half strength. If you want to use daylight or other light sources, you could use heavy shade cloth to exclude some of the light in a greenhouse.

When the lights come back on, the temperature can also go down to a more normal growing temperature, roughly 72 to 77°F (22–25°C). If you're using artificial light, make sure the lights aren't so close to the domes that they cause them to heat up excessively. When the plants are in a low-light environment, you also need to make sure that the light doesn't come from only one direction, since this will cause the plants to lean toward the light. I learned the hard way that newly grafted plants leaning toward the light will pull their graft unions apart as they bend.

Keep the plants in this low-light, medium-temperature, high-humidity environment for three

Figure 8.16. Grafted tomato plants that have healed on a light cart with humidity domes removed.

Figure 8.17. This photo comes from a commercial grafting operation. Recently grafted plants are on the bench tops. Entire benches can be made into healing chambers. Plastic can be rolled over the metal supports to make a tent to keep the humidity high.

If the Plants Aren't Healing Properly

Grafted plants cannot stay in the healing chamber forever. They have to be healed more or less in the time frame listed in this chapter in order to continue developing normally. Plants that are in healing too long may develop adventitious roots from the scion, begin to rot, and become peaked. So if you keep trying to lower the humidity and the plants wilt every time so you repeatedly have to back up, you may want to pull some of them apart and try to diagnose what went wrong (see "Troubleshooting" on page 136).

more days. Monitor the healing chamber to make sure humidity stays high, but otherwise don't change the conditions. After three days in this transitional environment, make a very small opening in the healing chamber to release a little bit of humidity. You want to begin the transition back to normal growing conditions by starting with the smallest of changes. You are gauging the ability of the plants to deal with any change, but without causing much stress in case they're not ready to transition.

If you make a small opening and the plants start to wilt, rehumidify the chamber, close it back up, and try again tomorrow. If you make a small opening one day and the plants do not wilt, you know you can move on to reducing the humidity even more the next day. The principle you can use throughout the healing process is: If you make a change and the plants don't wilt, you can move on to the next phase of healing on the following day.

If you make a change and the plants wilt, go back a step from wherever you are in the process. The plants may not be ready for the next phase, or it may be that you made too large a change for the plants' stage of healing. Backing up and rehumidifying will usually solve the problem and give the plants another day of healing to be able to adapt to less humid conditions the next day.

When you start opening your healing chamber, try to make it have an even effect on the humidity inside. What you don't want is for the humidity to plummet in one area of the chamber (close to the opening), while it stays high in the area far away from the opening. This is a bigger problem with long, skinny healing chambers, like those constructed on a long greenhouse bench or rectangular table. Opening up one end may have little effect on the humidity at the other end. The ideal situation would be to make a very small opening the length of the chamber, so all the plants experience the same amount of reduction in humidity. This is less important with very small healing chambers, like humidity domes. The easiest domes to start ventilating are those with adjustable vents on the top and sides. On the first day you want to start reducing humidity, you can simply make a tiny opening in the top vent, and proceed to larger openings on subsequent days.

If you're using humidity domes that don't have pre-made vents, you can use something to prop one of the long edges up a tiny bit and start releasing some humidity. Opening a long edge will provide a more even reduction in humidity than opening an end. Make the first decrease in humidity very small. If the plants are not ready yet and you drop the humidity a lot, this will stress the plants and hamper their recovery.

Your goal is to gradually reduce the humidity over the course of three or four days until all the vents are open or the chamber is back to ambient humidity. If the plants are not wilting that means the plants are maintaining turgor pressure through transpiration with moisture supplied through the roots again. If plants need to be watered at this point, continue using bottom water. At some point the graft union will be as strong as an uncut stem, but give them some time to heal before top watering again.

After a couple of days with reduced light, reacclimate the plants to normal lighting, especially if you notice them stretching under the low-light conditions. Once they have spent a day at normal light, temperatures, and humidity, they are ready to go back to greenhouse conditions. Try to put them back in the greenhouse on a mild day, early in the morning before it's hot, or reduce the airflow, because it is still a step in the acclimation process and you want to make it a smooth one.

Check on the plants an hour or two after they go back in the greenhouse just to be sure they're making the transition smoothly. Don't assume at any point that just because healing is going as planned, you can skip ahead more than one step at a time. I have had batches that were doing well until I rushed them and then they died. Even perfect grafts healed under ideal conditions need to transition gradually.

This is a good time to add support sticks if you are using them. The plants will start growing quickly again and become prone to falling over. Poke them through the pocket in the grafting clip and into the medium. If you didn't use clips with a pocket for support sticks, you can rubber-band them to a stake as described in the seedling section.

Handling Grafted Plants

Once the plants have been healed and reacclimated to greenhouse conditions, there are only a few small differences you need to keep in mind when handling grafted seedlings. When it comes time to transplant, you want to keep the graft union above the soil line, ideally by an inch (2.5 cm) or more to keep the top variety from forming adventitious roots.

Keep an eye out for any suckers that form from below the graft union, since they will be the rootstock variety and not produce usable fruit. If you used silicone grafting clips or clips made out of surgical tubing or some other material, these will be pushed off the stem as the plant

Figure 8.18. Trays of grafted plants with sticks for support. The sticks fit in a hole in the grafting clip.

Figure 8.19. A close-up of the sticks used to support grafted plants using the grafting clip as an anchor.

Figure 8.20. Here is a tray of grafted tomato seedlings that have healed and been placed back in the propagation greenhouse.

Figure 8.21. Start with a seedling like this for cleft grafting. Photo courtesy of Azad Photo.

grows. It is not necessary to remove these. It may be necessary to remove spring-loaded grafting clips. Other than that, grafted seedlings can be handled as normal.

Top-Grafting Troubleshooting: Mismatched Stems

If you go to top graft and your stem diameters are poorly matched, there is a modified top-grafting method you can try. Or if the stems are so poorly matched that they can't be top grafted, there are other grafting methods that are more forgiving of mismatched stem diameter (side grafting and cleft grafting are discussed below).

The closer your stems are to matching, the better your top grafting will be. The only way to know how mismatched is too mismatched is to try some poorly matched grafts and see what you can get to heal successfully. This is why it's so important to learn on expendable plants.

If your top and bottom varieties are close but do not match (the tops are smaller than the bottoms or vice

Pinched-Top Grafted Plant Production

One variation on top-grafted seedlings is to produce what's known as pinched-top seedlings. This type of plant is standard in the greenhouse industry, though I hardly ever see it outside big commercial greenhouses.

Pinched-top seedlings are used to make a double-leader plant where both vines are as evenly matched as possible. It's particularly important when lowering and leaning tomatoes to have the stems evenly matched, so they are all ready to be lowered at the same time. I have always just used the sucker below the first flower cluster to develop a second head; this technique has produced stems that are similar enough for my needs. But if you've had problems with your leaders being poorly matched, you could consider this technique.

Pinched-top plants are made from a top-grafted seedling. Once the seedlings are healed and have started to grow again, cut the top off the plant just above the cotyledons (the cotyledons of the scion). Though it may seem like this would kill your newly grafted plant, the removal of the top activates latent buds that are present at each cotyledon. In normal growth, these buds do not develop, because the main stem is dominant. But when you remove the top, the plant begins to grow from these dormant buds in a last-ditch effort to grow.

This is an extra hassle to go through in producing the seedlings and adds another two weeks to the production time. So seed that much earlier if you plan on making pinched-top plants.

versa), you can change the placement of the cut in relation to the cotyledons in order to find a place on the stem that is a better match. You can take advantage of the fact that the stem is smaller above the cotyledons than it is below. For example, if your rootstock is larger than your top variety, you can make the cut above the cotyledon on the rootstock to get a smaller section of stem. Make the cut in the normal place (below the cotyledon) on the scion variety; this may match better than making the cut below the cotyledon on both varieties.

This can usually only be done on the section of stem between the cotyledon and the first true leaf. The stem is almost perfectly round below the cotyledon, and it tends to get more oval the farther up the plant you go. So if your plants are really mismatched, the size match may improve but the shape of the stem will become harder to match farther up the plant.

The disadvantage of grafting above the cotyledons on the rootstock is that the rootstock may grow from the cotyledons. This will produce a seedling with two sets of cotyledons – from the rootstock and from the scion. The same potential for latent bud growth that makes pinched-top seedling production possible also makes regrowth likely when you leave the cotyledons on the rootstock. This is a nuisance when it occurs because suckers that grow from the rootstock will compete with the main stem. Since they are coming out of the rootstock, they won't produce salable fruit. If you notice suckers coming from anywhere on the rootstock, they should be removed for this reason.

Likewise, if your scions are larger than your rootstocks, you can make the cut on the scion above the cotyledon to try to achieve a smaller stem diameter to match the rootstock when cut in the normal place below the cotyledons. There are no cotyledons on this type of plant at all, which does not permit pinched-top seedling production. So it's not really a problem unless you were planning on doing pinched-top production.

You should still plan on trying to produce seedlings that are as closely matched as possible, because this is the easiest way to graft. But this can be a useful trick to deal with plants that don't quite match. I have used this technique when necessary with a high rate of grafting success and very little suckering from the cotyledons. But the amount of suckering from the rootstock depends on variety and greenhouse conditions. So always keep an eye out for suckers below the graft union with this method.

Solanaceous Cleft Grafting

As noted above, cleft grafting is more forgiving of differences in stem size than top grafting. It's also called wedge grafting, because it involves inserting a wedge-shaped scion into a vertical slit in the rootstock to splice the two plants together. If one stem is larger than the other, cleft grafting demands that it be the rootstock.

The downside of cleft grafting is that it takes longer than top grafting, because more cuts have to be made for this type of splice. Cleft grafting is as unforgiving of poor healing conditions as top grafting, since the top is completely severed from the roots. So this is not the method to use if you have less-than-ideal healing conditions. I mainly recommend this method as a fallback in the event that your rootstocks are larger than your scions.

Plants are prepared in the same manner as for top grafting, but it may be easier to do cleft grafting on plants that are larger than the ideal for top grafting. Instead of cutting the rootstock off at an angle, for cleft grafting the top is removed from the rootstock with a cut straight across the stem, just below the cotyledons. Then cut down into the rootstock roughly 0.125 inch (4 mm), splitting it in half from the cut down.

There are two methods to prepare a scion for a cleft graft. Scions for cleft grafting can be cut above or below the cotyledons, depending on what matches better. The first method is to cut the scion off straight across. Then make a cut starting 0.125 inch (4 mm) above the stump, angled approximately 65 degrees into the stem, ending close to the middle of the stem, eliminating a wedge about a third of the diameter of the stem. Make another identical cut on the other side of the stem, so that instead of a flat cut the stub is now a wedge.

The other way to cut the scion for a wedge graft is to cut the wedge out of the seedling without decapitating it first. Make a 0.125-inch (4 mm) cut, angling toward the middle of the stem, forming one side of the wedge. Then make the same cut on the other side, with the second cut forming the other side of the wedge and severing the top. The result is the same as the first method for making a top for cleft grafting, but it's faster because you eliminate the step of severing the top from the roots before making the wedge-shaped cut.

Put a silicone clip over the rootstock stub, and slide it down below the slit in the rootstock so it's out of the way before you add the scion. Then insert the wedge

Figure 8.22. Decapitate the rootstock seedling just below the cotyledons and split the stem in two. Photo courtesy of Azad Photo.

Figure 8.23. Here is the scion being fitted into the cleft in the rootstock. Photo courtesy of Azad Photo.

Figure 8.24. Here is the scion in the cleft in the rootstock. Note that the grafting clip is below the cut, ready to be slid into place. Photo courtesy of Azad Photo.

Figure 8.25. The finished graft with the clip moved into place. Photo courtesy of Azad Photo.

at the base of the scion into the split in the rootstock. Slide the tube or grafting clip up around the graft union so that it keeps the wedge in place. Heal as with top grafting.

Solanaceous Side/Approach Grafting

The side-grafting method, also known as approach grafting, is the most forgiving both of differences in stem size and of less-than-ideal healing conditions (see figures 8.27 through 8.29). This is because the roots of both plants remain attached until the graft is healed, so the top has a support system while it's healing. The connection to the scion's roots is severed only after the graft has healed.

The downside of side grafting is that it's more consumptive of time, materials, and space than the other methods. That's why I recommend it as a fallback method when stem sizes have become too mismatched to graft any other way.

Side grafting starts off like the other methods, by planting batches of the desired rootstock and scion varieties. As in the other methods, the closer the stem sizes are to matching, the better the results, but side grafting can overcome some fairly big differences in scion and rootstock stem size. Only experience can tell you how much you can get away with. If you have plants you intended to top graft that have gotten too big or mismatched, you can try to save them by side grafting if starting over would throw your timing off.

You do need to have some different supplies on hand for side grafting than for the other methods. The larger stem sizes and the fact that the clip has to go around both stems mean that larger, spring-loaded grafting clips work better for side grafting. You also need pots large enough to hold both root balls.

When it comes time to side graft, pop the rootstock seedling out of its tray and cut the top off the rootstock variety just below the cotyledons. Find a scion seedling of as close to matching stem size as possible, pop it out

Figure 8.26. Large grafted tomato seedlings that are ready to be transplanted on the author's farm. Bamboo skewers have been inserted through the loop in the grafting clip, but they will need to be rubber-banded to the skewer before they are transplanted to keep them from flopping over.

Troubleshooting

Grafting is a complex process with a lot of steps where things can go wrong. I will go through some of the most common mistakes I have made and witnessed to help you with your own troubleshooting. Hopefully, this will save you from making the same ones. Once you become proficient at grafting, success should be consistent when you can create the same conditions every time. Make notes and keep track of when you did what, so you can remember when to move from one stage of healing to the next, but also so that if things don't go as planned, you can remember what you did and do it differently the next time.

Plants Survive but Don't Thrive. There will be plants that survive the grafting process, but after a week or two of growth they lag behind the other grafted seedlings. This can happen when the vascular system heals enough for the plant to survive but not completely. Don't use these plants; they will never grow as well as the plants that have healed properly. This is why it's important to overseed for grafting more than you would for the field; since you have a chance to lose plants at germination and at grafting, you want to make sure that you can leave out any bad plants and still have enough.

Plants Wilt Quickly After Grafting. If the plants are wilting very quickly after grafting, they may still be respiring too fast. Try to slow them down by keeping them in a cooler environment for longer before cutting. If the plants are wilting in the healing chamber, there may not be enough humidity. Or the roots may be too dry.

Grafts Are Not Joining. If the grafts have not joined, even after spending the right number of days in the healing chamber, it's possible that the cuts or stem diameters were not close enough to matching, or that the unions were knocked apart after grafting. This could happen if the plants were watered or handled too roughly, or exposed to wind that knocked the grafts apart just a little bit. Keep in mind that the stem cuts must be in contact with each other to heal together.

Adventitious Root Formation. Adventitious root formation from the scion can be a response to stress, or it can be a sign that the graft is not healing properly. Adventitious roots may also form if plants are kept in the high-humidity healing chamber for too long. Ultimately, these roots are not a problem on their own, as long as the graft union is healing properly. But if adventitious roots form from the scion, do not let them develop or they will be susceptible to soilborne diseases that the rootstock may be resistant to.

of its tray, and place it next to the decapitated rootstock seedling. Next, make a cut in the stem of both the rootstock and the scion, and join them together with this cut. Until you get the hang of it, it can be helpful to take the two stems between your thumb and forefinger and gently pinch the stems together to see where they meet, close to the top of the severed rootstock. Where the stems meet when pinched together is where you want to place the cuts on each stem. If there are any leaves on the scion where the stems come together, it's a good idea to cut them off to get them out of the way. If you cut leaves off, make sure to cut them all the way back to the main stem; don't leave any stubs. Stubs tend to rot in the warm, high-humidity environment of the healing chamber.

Make a cut at a 65- to 70-degree angle downward into the rootstock, about two-thirds of the way through the stem. Then make a corresponding cut at a 65- to 70-degree angle upward into the scion, two-thirds of the way through that stem. It doesn't matter which stem gets the upward cut and which gets the downward, as long as they mate up when the time comes to put them together. Once you have made cuts in both stems, each will have a flap of tissue. Fit the two flaps together so the two stems are as fully integrated into each other as possible. Take a spring-loaded grafting clip and place it over the graft union so it holds both stems together.

In order to move the plant to the healing chamber, you have to place it in a pot large enough to hold both root balls. Pot it up as you normally would move a seedling from a smaller to a larger container. Put some soil in the bottom of a pot, place both seedlings in the pot, and put enough soil around the root balls to cover, making sure not to get soil in the graft union.

If you can make a healing chamber as described in the top-grafting section, that is ideal, but side grafts can heal under a much broader range of conditions than those necessary for top or cleft grafting. Many people who can't make a proper healing chamber are able to heal side-grafted plants under benches or in makeshift tunnels in greenhouses. I've been able to heal small numbers of plants in a cardboard box that I misted for humidity and placed in a greenhouse before reintroducing light. As long as you can increase humidity, decrease light, and keep the plants at a reasonable tomato-growing temperature, you have a good chance of being able to heal side-grafted plants. Test your setup for efficacy before using it on important plants, especially if it deviates much from the ideal healing chamber conditions. But you can get away with much less optimized conditions with side-grafted plants.

Approach grafts don't have to be in the dark immediately following grafting, so put them in a reduced-light, high-humidity healing chamber for three days. On the fourth day, take the plants out of the healing chamber and cut halfway through the scion variety's stem to begin weaning the plant off that root system. Remember that the scion is the one that still has its top. Make sure not to cut the rootstock, which doesn't have a top. Then place the plants back in the healing chamber for two more days. On the third day, remove the plants from the healing chamber and cut the rest of the way through the scion variety's stem so that the plant is relying only on the rootstock. The plants may be tall enough that they are floppy by this point. If so, rubber-band them to a bamboo skewer or other support to help hold them up. At this point the plants should be ready to go back in the greenhouse. Reintroduce them to the greenhouse in the evening or on an overcast day and check on the plants an hour or two afterward, to make sure they're able to deal with greenhouse conditions.

Cucurbit Grafting

Currently, most cucumber rootstocks are species other than the domesticated cucumber (*Cucumis sativus*). Other species are used because they may be resistant to diseases no cucumber is resistant to, and/or because they may have a larger, more vigorous root system. The most common cuke rootstocks at this point are species such as figleaf gourd (*Cucurbita ficifolia*) and butternut squash (*C. moschata*). Just like solanaceous rootstocks, interspecific crosses such as *Cucurbita maxima* x *C. moschata* (for example, the variety Tetsukabuto) are also being used to take advantage of interspecific hybrid vigor.

Unlike nightshade rootstocks, cucurbit rootstocks may have an influence over the look and flavor of the fruit produced by the scion. Some cucumber rootstocks can impart off flavors or more or less of a bloom (the waxy covering some cucumbers have) to the fruit. If you are going to graft cucumbers, make sure the rootstock does not have any unwanted influences on the top variety. You can confirm this with the seed company selling the rootstock, or with your own trials if they don't know. Rootstocks developed specifically for cucumbers should not have this problem, but other cucurbit species could show unpredictable results. This has been a particular problem with melon growers, where using the wrong rootstock imparts an undesirable squash or cucumber flavor, which makes the melons unmarketable.

At this point, cucumber grafting has not taken off in North America. Rootstock varieties have come on and off the market due to lack of interest. In my opinion, cucumber grafting has the potential to take off when rootstocks are developed that have both the increased disease resistance and yield boost seen in tomato grafting. Currently, disease resistance is the main advantage of cucumber rootstocks, and this is of great interest for cucumber growers with soil disease problems. But my guess is that it will take yield boosts like those seen in grafted tomatoes for cucumber grafting to catch on with those who don't have soilborne disease problems. I include this description of cucumber grafting for those who have disease problems, and in the anticipation that yield-boosting cucumber rootstocks may not be far off.

Cucurbit Side Grafting

Side grafting can be used on cucurbits much as on solanaceous crops described above. Try to time scion and rootstock so the stems are of matching diameters. When you're using rootstocks with larger seedlings than the scion, you may need to start the scion before the rootstock. A good example of this would be grafting a cucumber onto an interspecific squash rootstock.

If you're not sure, compare the seeds. A larger seed will usually produce a larger seedling. For example, most squash seeds are several times larger than cucumber seeds; you can be pretty sure they'll make larger seedlings. Do a test planting to figure out the exact timing depending on the varieties you're using. Or your seed source might have timing suggestions if they are familiar with both varieties. Though one of the advantages of approach grafting is that it's more forgiving of small differences in stem diameter, the closer you can get to matching, the better.

Otherwise, follow the directions for solanaceous approach grafting described above, using a spring-loaded

Figure 8.27. This is a good-sized seedling to approach graft, though the method is more forgiving of varying stem sizes. Photo courtesy of Azad Photo.

Figure 8.28. Make a downward cut two-thirds of the way through one plant. Photo courtesy of Azad Photo.

Figure 8.29. Make an upward cut on the second plant that matches the cut on the first where the two come together. It doesn't matter which you do first or second, but decapitate the rootstock just below the cotyledons (not shown) so there's no chance the rootstock will grow, and so there's no chance of confusing it with the scion. Mate the two cuts together and clip them in place with a spring-loaded grafting clip. Photo courtesy of Azad Photo.

grafting clip to hold the grafted plants together. Heal as described, though as with solanaceous approach grafting, this method is more forgiving than others since the roots stay attached until the plants are healed.

Cucurbit One-Cotyledon Grafting

Top grafting has a variable success rate with many cucurbits. One way to get around this is one-cotyledon grafting, the cucurbit equivalent of top grafting. When top grafted, cucurbits sometimes run out of energy before they can heal. The carbohydrates in the cotyledon left on the rootstock act as a source of energy to help the plant survive the grafting process.

The stems for one-cotyledon grafts need to be well matched, as with top grafting. See the method for producing matching seedlings in cucurbit approach grafting, above. Take your rootstock and cut at a 45-degree or greater angle between the cotyledons, severing one cotyledon and leaving the other. There is an axillary bud (tiny dormant shoot) that sits just on top of each cotyledon. It's important to cut very close to the remaining cotyledon to remove this bud along with the main stem and other cotyledon. If the axillary bud remains, it may begin to grow. Since it's from the rootstock, it would compete with the scion and produce unmarketable fruit. Then take your scion and cut at a matching angle 0.5 inch (10–15 mm) below the cotyledons. Put the two cut surfaces together and secure them with a spring-loaded grafting clip. Heal as described above. Cucurbits may need slightly longer in healing than solanaceous crops.

Figure 8.30. Begin with a cucumber plant like this for one-cotyledon grafting. Photo courtesy of Azad Photo.

Figure 8.31. Cut the top off the plant just above one cotyledon, cutting below the cotyledon on the other side. Make sure to remove the latent bud just above the cotyledon you're leaving or it may grow out, producing an undesirable shoot of the rootstock variety. Photo courtesy of Azad Photo.

Figure 8.32. Make a cut at the same angle just below the cotyledons of the scion variety. Photo courtesy of Azad Photo.

Figure 8.33. Put the two cuts together and hold in place with a spring-loaded grafting clip. Photo courtesy of Azad Photo.

CHAPTER NINE

Tomatoes

The tomato has been studied more than any other greenhouse crop, and the complexity of keeping it productive over a long greenhouse season has given many a horticulturalist job security. The tomato is also the most, or one of the most, profitable crops on many mixed-vegetable farms. So the importance of the crop cuts across many sizes and types of farms. The enormous size of the fresh tomato market means there will always be many growers of all different sizes competing in the market.

Figure 9.1. Tomatoes being grown aquaponically with coir as the medium (see appendix A for more about aquaponics).

Ideal Tomato Climate

OPTIMAL TWENTY-FOUR-HOUR AVERAGE TEMPERATURE

66°F (19°C)

MIN/MAX TEMPERATURE

Try to prevent temperatures over 86°F (30°C), or below 60°F (15°C). Temperatures lower than this range may cause flowers to develop abnormally, while temperatures below 48°F (9°C) will cause serious setbacks in growth. This may lead to catfacing, fused fruit (see below), zippering, and other abnormalities, especially in heirlooms or other varieties prone to misshapen fruit. Pollen may fail to develop in flowers exposed to below-ideal temperatures, which causes blossoms to drop without setting fruit. Temperatures above the ideal may cause pollen sterilization, which will also cause blossoms to drop.

CLIMATE

For the week after transplanting tomatoes to the production greenhouse, maintain a flat temperature profile of 73 to 77°F (23–25°C) both day and night to encourage rapid rooting in and vegetative growth.

After the first week in the production house, tomatoes can be grown from 75 to 76°F (24°C) during the day and 64 to 66°F (18–19°C) at night. When fully loaded with fruit, grow from 75 to 80°F (23–27°C) during the day and 64 to 67°F (17–19°C) at night. Pre-night temperatures down to 61°F (16°C) can be used to steer the plants in a more generative direction.

Keeping temperatures below 86°F (30°C) can be a challenge in the summertime. I can say from experience that under many circumstances, tomatoes can withstand temperatures hotter than that without compromising pollination and without major damage to the crop. However, flowers may be damaged by temperatures over 86°F (30°C). Under adverse conditions in heat waves I have seen greenhouses go into the triple digits F (over 37°C) for a few days without major adverse effects. On the other hand, at temperatures above 96°F (36°C), respiration outpaces photosynthesis. This means that the plant is using energy faster than it can produce it, so periods above this temperature need to be minimized (see "Respiration" in chapter 5).

If you know you're in for adverse weather conditions, keeping the structure cool for at least the morning can help keep flowers from aborting. If the temperature goes above the mid-90s F (low 30s C) before noon, you run the risk of sterilizing flowers that opened that morning. If you live in a hot climate and plan on producing tomatoes through the summer, using a vigorous variety or grafting to a vigorous rootstock can help. Since prolonged excessively hot weather has the effect of weakening

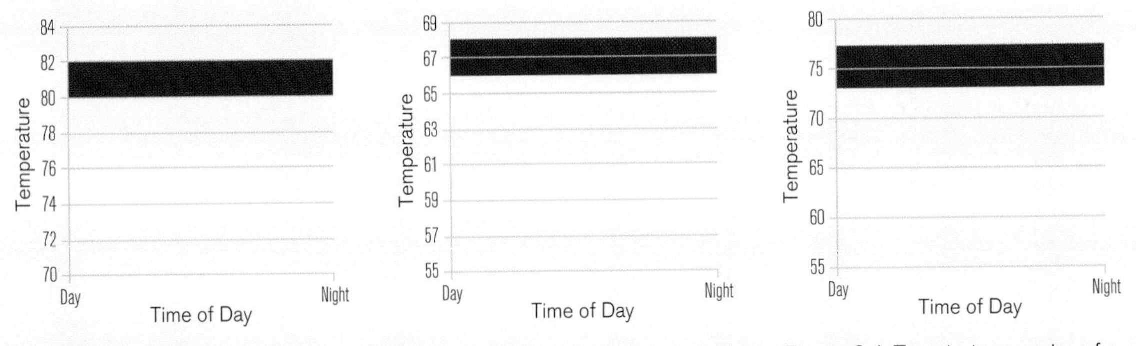

Figure 9.2. Tomato germination temperature.

Figure 9.3. Tomato seedling growth temperature.

Figure 9.4. Tomato temperature for the week after transplant.

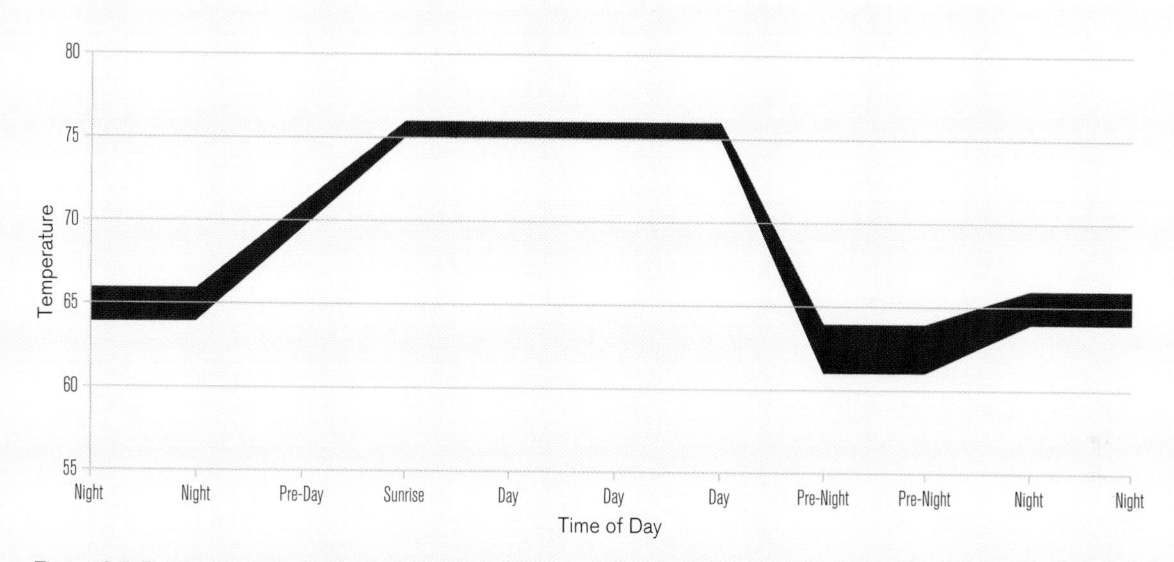

Figure 9.5. Tomato temperature at the beginning of harvest.

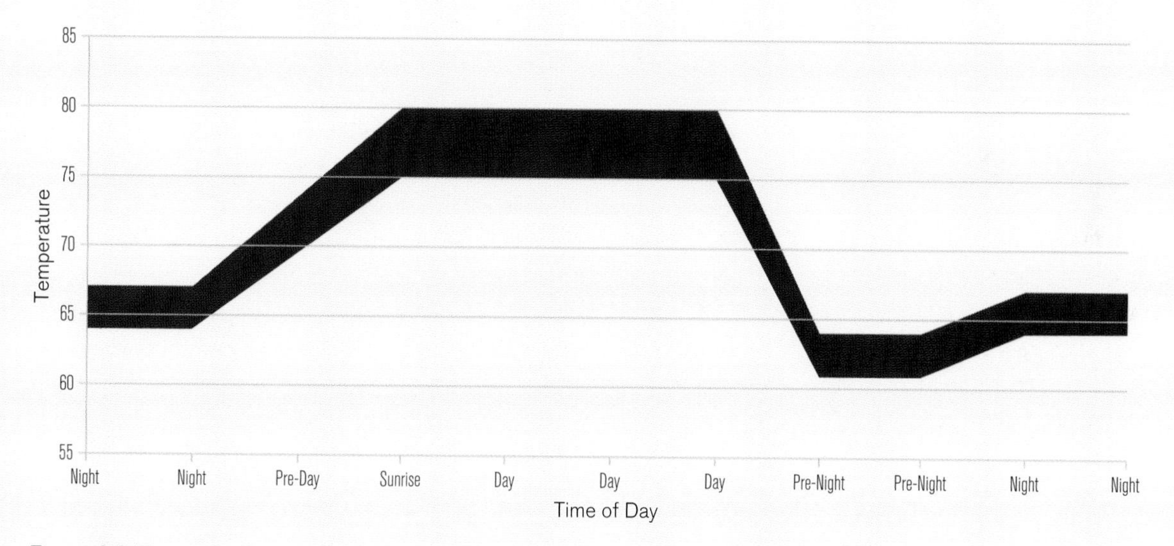

Figure 9.6. Tomato temperature at fully loaded harvest.

tomato plants, starting with a strong variety can help mitigate the effects of heat on the crop.

MEDIUM TEMPERATURE

The temperature of the roots is an important consideration in a tomato crop. The ideal is from 64 to 70° F (18–21°C). Root temperature is a limiting factor for in-ground heat; this is the reason you can't crank radiant heating up higher than the above temperature.

Low root-zone temperature can be a problem, especially for soil growers in cold climates. When a piece of plastic or glass is all that separates your greenhouse from freezing weather outside, that cold may infiltrate through the soil under the wall of the greenhouse into the root zone. If the root-zone temperature goes below 61°F (16°C), serious reductions in growth may become apparent. This is why some growers install frost walls.

Varieties

There is more specialized breeding for tomatoes than for any other protected crop. The importance of tomatoes to greenhouse producers and the specialized adaptations that separate them from field tomatoes have led to thousands of varieties bred specifically for protected culture.

One of the most fundamental ways in which greenhouse tomatoes differ from field varieties is in the disease-resistance packages. Many greenhouse tomatoes have the basic resistances that field tomatoes have. Resistance to fusarium, verticillium, and tomato mosaic virus can be useful in both field and greenhouse growing. But there is a group of diseases that are rarely found in field growing, which are much more common in protected culture. Resistance to diseases such as leaf mold, fusarium crown and root rot, and powdery mildew are the hallmarks of greenhouse tomato disease packages.

When choosing tomato varieties, it's important to know which diseases are likely to affect you. Unfortunately this knowledge usually comes the hard way, from experiencing disease pressure. If you have never done protected growing in your area before, ask your neighbors what diseases they have to deal with.

Though simply being grown under cover may eliminate or alleviate the pressure from certain diseases, it's usually a trade-off for other diseases more common in protected culture. Leaf mold, botrytis, and powdery mildew are particularly common on tomatoes in poorly ventilated houses, because they thrive in warm, high-humidity environments. This is where the increased climate control of a greenhouse plays a part in reducing disease. Keeping the temperature and humidity in the ideal range will reduce the incidence of these pathogens. And scourges of field tomato production, like early and late blight, are usually not even included in greenhouse tomato breeding programs. Greenhouse varieties are at a much lower risk for these diseases, since the plants don't get rained on, and soil splash and moisture on the leaves are the factors that predispose plants to early and late blight infection.

Other features of greenhouse tomato breeding may include a high level of vigor to keep plants strong over a long season, as well as plant openness to maximize airflow in densely planted situations.

At this point, almost all of the breeding for greenhouse tomatoes is focused on indeterminates, because most of the market is for greenhouses that want a very long harvest season. Since most determinates do not bear for as long as indeterminates can, they would require multiple plantings a year and increase the amount of establishment time. Also, the flavor of determinate tomatoes tends not to be quite as good as indeterminates, partially because indeterminate tomato varieties have three leaves between each fruiting cluster and determinates have only two. Determinates have a lower ratio of leaves to fruit — in other words, fewer solar panels for the sugar factory.

A lot of growers like determinate tomatoes, however, because they require less labor, so they compromise and use semi-determinate tomatoes to combine that benefit with the ability to still stretch the season longer than true determinates. Some of these varieties can get quite tall in protected culture. They grow taller than they typically would in the field under the lower-light conditions inside a structure.

Shorter-season determinates can be attractive to some growers, especially in hot areas. They may sneak in a quick crop of determinate tomatoes before or after the field tomato season, and then leave the hoophouse unplanted during the heat of the summer when it's really too hot inside.

The downside of all this breeding is that good shelf life and high production tend to be prioritized, while flavor takes a backseat. These breeding goals have yielded many varieties that direct-market growers find less than inspiring. Since market farmers have a shorter supply chain, shelf life is not as important as for the big growers. And since more often than not, their farm's name is associated directly with their produce, they make their reputation (and draw repeat customers) by having the tastiest tomato at the market.

It's not that yield and shelf life don't matter for smaller growers, it's just that the priority ranking of these traits differs from growers who have to ship longer distances (and the emphasis on everything except flavor is not unique to greenhouse tomato breeding; it's the same for most field crops as well).

In the past these breeding priorities resulted in very hard tomatoes, because a firmer tomato is going to handle better and have a longer shelf life than a softer one. Lately, though, it seems like more attention is being paid to flavor in greenhouse tomato breeding. I believe this is because the main advantage that greenhouse tomatoes can claim in the grocery store over field tomatoes is better quality. The produce has to live up to that promise if it's going to keep its market share.

I can affirm that the quality of greenhouse tomatoes is higher now than when I first started doing trials and looking at a lot of varieties. New varieties are being developed that incorporate better flavor along with the durability improvements of contemporary breeding. This makes it more likely that growers will be able to find varieties that meet both their production and their flavor goals.

Tomato Variety Types

The huge importance of tomatoes to greenhouse vegetable growing has resulted in an amazing diversity of cultivars with specific greenhouse adaptations. There are varieties of almost any type you might want to grow available for protected growing. There are even a couple of types – the cluster tomato and truss cherry – that are specific to greenhouse culture.

Jointless varieties, for example, represent a modern innovation for cluster tomatoes. Jointless tomatoes lack the abscission layer where an individual tomato can be removed from the truss. This means that they do not fall off the cluster as easily when being handled. This trait is also sometimes used in field tomatoes so the picked tomatoes do not retain their calyxes, which can puncture other tomatoes during harvest.

Though the beefsteak was the mainstay of the greenhouse tomato industry for a long time, smaller varieties have evolved alongside demand for snacking vegetables. If you told growers twenty years ago that grape, cocktail, and cluster tomatoes would be important categories in the near future, they probably wouldn't have believed you – if they even knew what those types were.

I will run through the different greenhouse tomato types from smallest to largest. Keep in mind that smaller-fruited varieties will be lower yielding, more or less in proportion to their size (the smaller the fruit, the lower the yield).

Grape Tomatoes

True grape tomatoes are not just oval-shaped cherry tomatoes. Grape tomatoes have a flavor and texture profile all their own due to being firmer and having less gel and seeds than cherry tomatoes. They are also usually less prone to cracking than cherry tomatoes. Even people who do not appreciate firm tomatoes may like grape tomatoes as long as they are firm without being hard. Good grape tomatoes are one of my personal favorites for eating out of hand.

Grape tomatoes tend to have a very open, wispy-leaved plant habit that is conducive to airflow. They are usually less affected by leaf mold than other types, though leaf mold resistance can still be useful where disease pressure is high.

Cherry Tomatoes and Truss Cherries

Like grape tomatoes, protected culture cherries have open plant types, along with low fruit loads and yields when compared with larger tomatoes. They tend to be sweeter and have better flavor than the larger types.

One subtype of the cherry tomato is known as the truss cherry. Instead of harvesting individual fruits, you harvest the entire truss of tomatoes by clipping it off the plant. Truss cherries are pretty much exclusive to protected cultivation due to the difficulty of producing a perfectly ripened truss of cherry tomatoes in the field. There is usually a target number of fruit on the truss, depending on how many can be fully filled out. If there are more fruits set on the truss than the plant can fill, you may need to nip the last few fruits off the tip off the truss so every fruit is perfectly filled out. The main feature that differentiates a cherry variety developed for truss harvest from one for loose harvest is that most truss cherries are jointless.

Important traits for truss cherry tomatoes include fast ripening and good shelf life, so the first fruit to ripen is not mush by the time the last fruit is ripe. These types are usually marketed in boxes to protect the delicate truss and fruit. This variety depends a lot on presentation, since the fact that it is on the truss shows that it was greenhouse-grown.

Cocktail Tomatoes

Cocktail tomatoes occupy a size class larger than cherry tomatoes and smaller than cluster tomatoes. There are varieties for loose harvest and jointless varieties for cluster harvest. This class of tomatoes is more popular in Europe than it is in North America. My impression is that North Americans like their tomatoes big or small but not in between, and this is an in-between variety.

Nonetheless, there is some market for this type of tomato, which can have very good flavor. It was originally popularized by a single variety, Campari, which is branded and sold under that name. It is sometimes marketed as a salad tomato, one that you can make two quick cuts to and throw on a salad.

Tomatoes on the Vine

Cluster tomatoes, known in the industry as tomatoes on the vine (TOVs), were bred specifically for the greenhouse industry. They have risen to take their place as one of the main greenhouse tomato variety types, along with beefsteaks. In a crowded produce marketplace, the greenhouse tomato industry was trying to differentiate itself from all the other tomatoes on the shelf, especially field tomatoes. The big idea was that if you saw these beautiful ripe tomatoes still attached to a perfectly clean vine that might even still have a whiff of tomato plant smell, the presentation would be so impressive it would set the tomatoes apart from others in the store.

And indeed it's very difficult to grow a perfect cluster of tomatoes in the field. The possibility of disease causing spots on a tomato, or rain causing split tomatoes, makes it much less likely that an entire cluster of tomatoes will make it out of the field in perfect condition. The other difficulty of producing cluster tomatoes in the field is getting them all to ripen close to one another. Even though TOVs are selected for fast ripening, having heat to keep the tomatoes ripening as quickly as possible is almost essential to getting the last one to ripen before the first one gets overripe.

Unfortunately, TOVs represent yet another example of produce whose best qualities — freshness and flavor — have been bred out. Many of the most commonly grown tomatoes in the industry recently have been varieties that prioritize production over flavor.

Still, some TOV varieties are much better than others, and have the potential to be quite good. Most varieties are bred to produce a cluster of four or five medium-sized or six smaller tomatoes. The technique is to grow the entire cluster until the all the tomatoes are ripening, and the one at the end (the last one to ripen) is at least starting to show some color.

Growers with a longer supply chain may pick the cluster with the end tomato still quite green, whereas growers with a shorter supply chain may be able to let the end tomato get riper. At that point the cluster is clipped off the plant and usually placed into the cardboard flat they will be shipped in, to minimize the amount of handling.

Some of the newer TOV varieties are jointless to minimize the amount of tomatoes that fall off the cluster. Many wholesalers have very stringent standards and will only buy clusters with a certain number of tomatoes. Shelf life and "green parts" (calyx and stem) that stay green for an extended period of time are important development criteria for this type.

As someone who has visited greenhouses with Dumpsters full of tomatoes outside, I have seen firsthand that length of shelf life is not just academic. If shelf life is a concern for you, you can do one of the most basic tests of shelf life yourself. Take several varieties at the same level of maturity, place them into boxes, hold them in conditions similar to what they will be stored at, and see which ones rot faster.

Beefsteaks

Beefsteak tomatoes used to be the main greenhouse type, originally produced to compete with out-of-season field production. The limitations of growing field varieties in greenhouses quickly led to the development of dedicated greenhouse varieties.

For people who are not familiar with the produce industry, I'm sure it sounds strange to say that one of the last things considered in the development of many new tomato varieties is flavor. But when you consider how many other breeding goals there are, you can understand how this has come to be, however unfortunate it is. Beefsteak varieties are now subject to the same rigors as other tomato varieties, namely to meet the highest standards of appearance, yield, and shelf life.

One way that shelf life is increased is by breeding firmer tomatoes, to the point where some of them are pretty hard even when completely ripe. There are a lot of beautiful greenhouse beefsteak tomatoes out there. Many of them look much better than they taste. Taste new varieties to make sure they are up to your flavor and texture standards before planting a lot of them.

One way that protected culture growers seek to differentiate their produce from field beefsteaks is to harvest with the calyxes on. Since most field tomatoes are harvested loose into bins, the calyxes are removed by the pickers so the stems don't puncture the other fruit. The fact that greenhouse tomatoes are usually harvested in a single layer straight into the flats they are shipped in means the calyxes can be left on without risk of puncturing other fruit.

If your harvest methods involve piling tomatoes, it's possible for pickers to clip the stems off and leave the calyx as they are harvesting. Even if consumers don't know what a calyx is, it does make for nice presentation to leave it on. Curved shears are useful for cutting the stem off flush with the top of the tomato.

Choosing Tomato Rootstock Varieties

The two main considerations that go into choosing a tomato rootstock are disease resistance and vigor. Disease resistance is fairly straightforward: Choose based on diseases you know are active in your soil. Even if you don't currently have active root diseases, I view grafting to a rootstock with a broad disease package as cheap insurance, because the way you find out if you have a pathogen in your crop is the bad way.

In the future, there may be more options for tomato rootstocks, including varieties developed especially for producing a particular fruit size or with a particular stress tolerance, such as tolerance of soil with a high level of salinity.

Tomatoes from Seed to Sale

Once you've chosen your varieties, it's time to think about how to make the most of them. Below are tomato-specific recommendations to take you from seed to sale.

Propagation

Deposit one seed per cell or plug in your desired medium. The ideal temperature for germinating tomato seeds is 80 to 82°F (27–28°C). Maintain this temperature as consistently as possible until germination. After emergence, maintain a flat day and night temperature of 66 to 68°F (19–20°C) for the next two to three weeks, until seedlings need to be transplanted into larger blocks or containers.

Grafting

See "Solanaceous Grafting" in chapter 8.

Spacing

There are a lot of different ways to space an indeterminate tomato crop. The basic principle is that smaller fruited varieties are spaced more closely, while larger tomatoes need more space. This is due in part to the fact that larger-fruited cultivars have larger leaves, which cast more shadow and take up more space. Smaller-fruited varieties have smaller leaves and more open plants, with grape tomatoes tending to have very sparse, open plants. With their relatively low yields, small-fruited varieties need to be packed in tighter to improve yields per square meter.

The other principle for greenhouse tomato spacing is that if you start very early in the year when light levels are low, you may want to start with a lower density and increase density by developing extra suckers into new plants as the light increases. For example, growers in the northern tier of the United States or southern Canada might transplant grape tomatoes in January or February at a spacing of 2.8 plants per square meter. A few weeks after transplanting to the greenhouse, when light levels are better, they might use a sucker to develop a second head on every grape tomato plant, giving them a final density of 5.6 heads/m^2. They might only develop a sucker into a new head on every other cherry tomato plant, for a final density of 4.2 heads/m^2. Only every eighth beefsteak plant might get a second head, for a final density of 3.15 heads/m^2. Each head is spaced out evenly along the wire as the plants are moved weekly for lowering and leaning. In this manner all of the heads can receive equal spacing even though not every plant has the same number of heads.

Density is calculated by the number of vines rather than per plant, since there may be more than one vine per plant. And densities may change throughout the season as growers add and remove heads; see above. So density refers to the number of heads, not the number of plants.

For determinate tomatoes, a double-row basket weave can be used to achieve similar or slightly lower densities than those listed above for the corresponding type of indeterminate tomatoes. Instead of referring to density by number of vines, the spacing for determinate tomatoes refers to actual planting density, since they aren't usually pruned to individual leaders. If a long season is desired from determinate tomatoes, you can use a semi-determinate variety like BHN-589, which may get quite tall, especially under lower light levels in protected culture. So use taller stakes than you would in the field. A starting recommendation for determinates would be a double row of tomatoes 2 feet (60 cm) apart, with a plant every 2 feet along each row.

Labor

Under the maintenance regime described in this chapter, it is estimated that indeterminate greenhouse tomatoes will require a minimum of 140 hours of labor to maintain per week per acre. This is with an efficient, experienced crew. Determinate tomatoes will require less labor since they don't require suckering or individual stem trellising.

Transplanting

In field growing, it's usually best to transplant tomato plants before the first flower cluster is open, because the plant is still very vegetative and will develop a large root system quickly. When a large transplant is produced, as is often the case in protected growing, often flowers will be open by the time of transplanting. When heat is used, the flat temperature profile in the week after transplanting will help the plant to be very vegetative and quickly develop a large root system, even though some of its energy is starting to go to flower production. Even in hoophouses, the improved climate will help the plant establish quickly even if it's starting to set fruit.

Otherwise, see chapter 6 for how to transplant tomatoes. Grafted tomato transplants should never be planted deeper than the original soil line. If the top variety contacts the soil, diseases can get into the nonresistant scion through ground contact.

Adventitious Roots

Tomatoes will easily form adventitious roots, or roots originating from the stem, where the stem comes in contact with the soil. Formation of adventitious roots in random areas where the stem is not in contact with the soil can be a sign of stress. It is common to take advantage of adventitious root formation in tomatoes to develop a larger root system at planting time. You can do this if your plants are not grafted. Bury the tomato seedling deeper than the soil line of the cell it is planted in, and roots will form where the stem is buried.

Be sure to plant grafted tomato plants with the graft union 1 inch (2.5 cm) or more above the soil line so adventitious rooting does not occur. If the top variety roots into the ground, diseases can get into the nonresistant scion through its roots. High humidity or debris around the base of the stem can cause rooting out as well.

Flowering

Flower initiation in tomato plants can begin shortly after the cotyledon stage. The cells that become the flowers start forming in the growing point long before they are visible. So stress at the seedling stage, especially low temperatures, can have a negative impact on the first fruit.

Temperatures below 60°F (15°C) interfere with normal blossom and fruit development and cause a variety of fruit deformities, such as catfacing, zippering, and an increased number of fused fruits. Catfacing is the presence of scarring and puckering at the blossom end of a tomato. Zippering is scarring that runs the length of a fruit and has a characteristic zipper-like pattern. Fused fruits occur when two fruits that would otherwise separate develop as one. This is most likely to happen in the fruit closest to the plant, known as the king fruit. Fused fruits are also sometimes called boat-shaped fruit, due to their characteristic long and curved shape.

All of these deformities can be due to stress early in flower development. One way you can tell a fruit is made from multiple fused flowers is because a single tomato flower or fruit usually has five sepals. If you see large irregularly shaped fruit with more than five sepals, you know that they are a product of multiple flowers fused together.

The cause of deformed fruit can be the source of some confusion, because growers think they're getting fruit deformities even though the flowers weren't exposed to cold temperatures. What they don't understand is that even though no flowers were visible, they may still have been present in the growing point in an undeveloped state when plants were transplanted out early or otherwise exposed to cold temperatures.

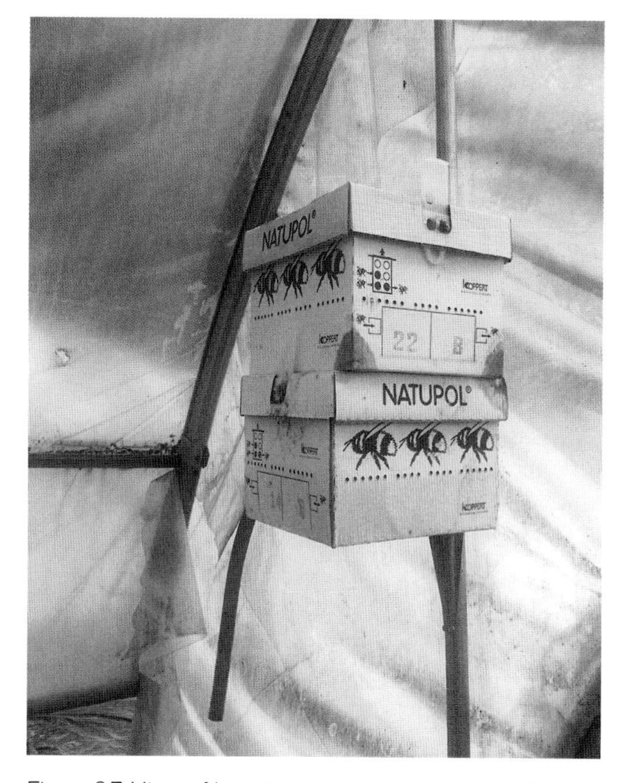

Figure 9.7. Hives of bees in a greenhouse for the pollination of tomatoes.

Pollination

Tomato flowers require pollination in order to set fruit. In the field, pollination occurs through the action of the weather and insect pollinators on the flowers. In small, open-sided hoophouses and greenhouses, wind and insects usually penetrate far enough into the structure for pollination to occur. Larger, multibay structures without rollup sides and those that have insect screening, however, may not have enough natural pollinators to set fruit, especially in the center of the house. In these cases pollination will have to be facilitated by the grower.

The most efficient and cost-effective option for pollination in greenhouses is to place hives of bumblebees in the structure and let them do it. In the early days of greenhouse growing, there were a variety of handheld vibrating pollinators being sold to get the job done. An electric toothbrush can even be used to pollinate. But experience has proven that it is more thorough and cost-effective to use bees, rather than to pay someone to go through the crop and vibrate the flowers.

A number of companies sell bumblebee hives specifically for the purpose of pollinating greenhouses. The vendor can advise you on how many hives you need to completely pollinate a house of your size. It is important to get the number of hives right: Not having enough may result in incomplete pollination and poor fruit set, while too many bees may actually "overwork" the flowers, which can damage them.

Bumblebees require a minimal amount of management and are not usually aggressive to people. Greenhouse hives also have some nice features – for instance, you can change the entrance so bees can enter but not exit. That way, if you need to spray or do some other activity that is not compatible with the bees, you can shut the exit before they come home for the night, and you know they will all be in the hive the next morning.

Trellising

The overwhelming majority of tomato varieties developed for protected culture are indeterminate. This is because greenhouses that want a tall, vining variety to take advantage of all the vertical space in a greenhouse and bear over a long season represent the largest market for greenhouse seeds. Those who want to grow determinates usually use a field variety with appropriate resistances for their area, grown in a basket weave or other field-style trellis system.

Determinate Tomatoes

Though a minority, there are plenty of determinate tomatoes grown in protected culture. In hot areas, growers may plant a crop of determinate tomatoes in the late winter or early spring and grow them as an early crop until field tomatoes are available. This is especially true where it gets too hot to grow tomatoes indoors through the summer. Depending on the climate and market, it could also be worthwhile to grow a quick crop of determinates late in the season to follow field tomatoes.

The advantage of determinates is that they can be basket-weaved to save time on trellising, and require little or no pruning. Some growers don't prune determinates at all, while others will remove a few suckers at the base. This can help increase airflow around the bottom of the plant and allow the plant to get bigger before it starts putting energy into fruit production.

The highest you normally want to prune a determinate is up to the sucker below the first flower cluster. Since determinates have a certain predetermined size, more pruning will begin to reduce the plant's limited opportunities for fruit setting.

Figure 9.8. The view along the pathway of a modern hydroponic tomato greenhouse. The blue string on the left was used to denote a sucker that was developed into a new head to increase planting density as light increased in the spring.

Indeterminate Tomatoes

The most common way of growing indeterminate tomatoes in protected culture is to grow each vine up its own individual support. Twine is installed above each vine before it begins to flop to give the plant support as it grows. Twine may be tied to overhead pipes or wires, or attached using movable spools that permit lowering and leaning as the plant reaches the height of the wire (see "Lowering and Leaning" in chapter 6).

I've occasionally seen indeterminate tomatoes grown in a tall basket weave in simple hoophouses. The reason you'd do this would be the same reason you'd grow determinate tomatoes: to save labor. Training tomato plants up a string, as is done in most greenhouse growing, is labor-intensive, but it is also facilitates a very high and regular yield of top-quality tomatoes.

Treating indeterminate tomatoes like determinates and growing in a basket weave, especially without pruning, will sacrifice some fruit size. The plant will have so much fruit set that no one fruit will receive nearly as many resources from the plant as on a pruned indeterminate. There will also be more second-quality fruit that develops. But if labor is short, it's an option.

The vast majority of tomatoes grown in protected culture are limited to a set number of stems, also called vines or leaders. The number of leaders is usually one or two per plant, but depending on variety and growing conditions sometimes more are used. The number of stems may even change throughout the year depending on light levels and growing conditions (see "Spacing" on page 147).

Pruning: How Many Stems per Plant?

Most greenhouse tomato production uses indeterminate plants pruned to one or two vines. The choice of how many leaders to use is made by taking into account the vigor of the plant and the duration of cultivation. The goal is to create a balanced plant that will have enough vigor to last the length of the season without being too vigorous.

Every leader on a plant takes a share of the plant's energy. Developing more leaders can be a way to balance the energy of an overly vigorous plant. On the other hand, using only one stem is a way to consolidate and maintain the vigor of a weak variety.

For example, most ungrafted plants are grown as a single leader, because this drives all the plant's energy up through one stem, making the plant as strong as possible. Most grafted plants are grown with two leaders, because most vigorous rootstocks would be too vigorous if grown with a single vine.

Pruning a tomato plant is a matter of removing the shoots, or suckers, that develop at every node of the plant where a leaf is attached, leaving the desired number of stems. If suckers are left unpruned, each one will form its own vine. Each additional vine will compete with the main stem. To develop a plant with only one main stem is simple: You remove all the suckers that develop over the course of plant growth.

With a pinched-top plant, the process is the same but repeated twice for each of the stems that emerge out of the dormant buds below the cotyledons. Prevent any suckers from developing on either of the stems and you should get an evenly matched two-stemmed plant. This method is common in the industry but not often used on smaller farms.

The other way to make a two-stemmed plant, which I have always used, is to allow the sucker below the first flower cluster to develop into the second leader. This sucker is typically the strongest one on the plant, and has the best chance to make a second stem that is a good match for the primary leader. If this sucker is damaged or accidentally removed, any other stem on the plant can be developed into an additional leader, though it may lag behind the primary stem or take a while to catch up to it.

Under some circumstances, you may want to develop a plant with more than two heads. This is sometimes done when heading into the summer with high light levels, which can support a higher plant density than the low light of winter. Depending on variety and target density, every plant may not get a third head. Instead every third or fourth plant will get a third head, just enough to get the plant population up to the target density (see "Spacing" on page 147).

Some growers of grape and other very small-fruited tomatoes will grow plants with three or more heads. In this case, any sucker can be developed into an additional head. Just make sure to choose a sucker that is close to the top of the plant, because suckers that are down low in the shade won't grow very fast. This may be a strategy to reduce seed costs in varieties that require a high planting density by developing more heads instead of planting more plants. Results can be mixed using plants with four or more heads; it's worth experimenting with this before doing a lot of it.

Even though small-fruited tomatoes have a lower fruit load, if you put enough stems on a plant it will start to slow the plant down. It would be worth growing a row of plants with two or three heads, and a row with four or more heads, and comparing the yield. If the multiheaded plants result in more loss of yield per area than money saved in seed cost, then it is not a money-making strategy.

Figure 9.9. This is the kind of growth abnormality that heirlooms are prone to, but has been mostly bred out of hybrids. Note how the main stem ends in a flower cluster instead of continuing on with the vine. This plant can be kept going by developing the sucker below the dead end into a new head.

How to Prune

Pruning is usually a weekly job with tomatoes. Once you have figured out how many leaders you want the plant to have, pruning is a simple process of identifying any suckers that you don't want to develop into additional leaders and snapping them off by hand. Using your fingers is preferred because it is faster and may actually transmit less disease than cutting. Since the blade comes into contact with the interior of each plant, a disease in one plant could be transmitted to all the others.

It's ideal to remove suckers when they are about 1 inch (2.5 cm) long. They snap off cleanly from the plant, leaving a small wound that heals quickly. Suckers much smaller than an inch can be hard to grab. When suckers cannot be grasped properly, they sometimes get mashed instead of snapping off. When suckers are mashed on a first failed attempt at removal, they may need to be cut off later, which will take more labor and may create a larger wound. If suckers are missed and get really large, though, it's often better to cut them off with a blade or clippers; they become too hard to snap off cleanly when they get big.

You never want to remove all the suckers from the top of an indeterminate tomato plant until it's time to top the plant. You always want to have a few small suckers up close to the head in case something happens to the head, so you always have a sucker that can be developed into a new head. This is another good reason not to try to remove suckers much smaller than an inch (2.5 cm), because suckers that small will be up very close to the head of the plant, and should be left as potential head replacements anyway.

Things that could cause you to lose the head of the plant include a worker accidentally snapping it, or damage from pests or disease. Occasionally a plant will grow a "blind" head that terminates without a growing point. Greenhouse varieties have been bred not to do this; you will notice more blind heads and other irregular growth habits in heirlooms and other varieties that have had less breeding work.

Figure 9.10. The sucker below the flower cluster in this picture is starting to get big, but it's still manageable for easy removal by hand.

Pests, Diseases and Disorders

See thrips, whiteflies, tomato hornworm, and TMV in appendix B.

Blossom End Rot

Blossom end rot (BER) is a common disorder in tomato fruit when not enough calcium is reaching the end of developing fruit. It can produce a range of symptoms from dark markings under the skin on the side or end of fruit, to open, leathery spots at the blossom end. These leathery spots are susceptible to secondary infection by botrytis.

BER may be caused by inadequate calcium in the soil, uneven watering, or root damage. It is more prone to occur in excessively vegetative plants. When the leaves are growing very strongly, they can have a much greater pull on the nutrients in the plant than the fruit, and the fruit may simply not receive enough of the available calcium. Because of this, BER is common at the beginning of the season when plants are very vegetative. Under these conditions, blossom end rot can occur even though there may be plenty of calcium in the soil. Using generative steering to make the plant less vegetative may help.

Some varieties are more susceptible to BER than others. Elongated-shaped fruits, like Romas and San Marzanos, are more susceptible than round fruits. Fruits affected by BER should be removed as soon as you notice them. They cannot recover once they start showing symptoms.

Gray Wall, Blotchy Ripening, Yellow Shoulder, Internal Whitening

These are different names for a related group of physiological disorders in which portions of the fruit walls or core stay hard and flavorless and never ripen and color properly. The problem may be visible from the exterior, or it may only be apparent once the fruit is cut open. These disorders are similar to BER in that they can result from a nutrient deficiency in the root zone, and/or from environmental conditions. Though this does occur in field tomatoes, it's more common in protected culture.

This group of disorders is caused by a lack of potassium reaching the fruit. Adequate potassium is very important for proper ripening and flavor in tomatoes. So if the soil is deficient and the plant isn't receiving enough potassium in the first place, that can be the cause. Growers should consider applying additional potassium through fertigation or side dressing. Tomato plants that are ripening fruit need a lot of potassium, more than can be provided with a single up-front application.

Even if the plant is receiving enough potassium, there are several environmental reasons why adequate amounts may not be making it to the fruit. And just to confuse things, they can result from widely diverse conditions. Prolonged periods of cool, overcast weather can be the culprit, because the plant transpires slowly and may not be able to bring in enough potassium to meet the needs of a high fruit load. It can also be caused by hot weather, especially following periods of cool weather. This kind of weather pattern may occur in the summer when several overcast days give way to a stretch of sunny weather.

The hot-weather manifestation may be related to a decrease in root production during cloudy weather; roots are then unable to meet the plant's needs when this is followed by hot, sunny weather. One of the reasons this may be more common in protected culture is that temperature fluctuations are often more extreme. When the summer sun comes out after a few cloudy days, it gets hotter under cover than it does in the field. As with BER, excessively vegetative plants without much fruit load may be prone to this as well, as the rampant leaf growth pulls more strongly on the nutrients in the plant and hogs the potassium.

Botrytis

See more on botrytis in appendix B. Overfertilized plants with lush, soft growth are more susceptible to botrytis than balanced plants are. Temperatures below 65°F (18°C), poor airflow, and poor pruning techniques that leave stubs are conducive to developing this disease. Avoid all these conditions as much as possible.

Yield

The questions surrounding yield – what it should be, what it could be, how to maximize it, whether it comes at a loss of flavor – are tricky because they are subjective. One grower may have wildly different results than another with the same variety, because everything about how a crop is grown has an effect on the yield.

Yields of over 100 kg/m^2 of tomatoes per year have been recorded, but I don't think such yields are possible without a loss of flavor. They're only achievable when a maximum amount of water is pumped into the fruit, which of course dilutes whatever sugars and flavor compounds are in it. Once again we see how the interests of production sometimes conflict with the best interests of flavor. A more reasonable upper yield figure is probably in the 60 kg/m^2/yr (14.6 pound/ft^2/yr) for heated year-round production.

The best way to maximize yield and flavor is to optimize all the plant growing conditions, so that there are as few limitations to plant health as possible, without using techniques like overwatering to pump up the harvest. A plant that is healthy and growing well will be producing as much as possible and be able to devote a maximum amount of resources to production of fruit and flavor compounds. This is in contrast with a sick plant that has limited resources to devote to either end.

Genetics also play a role in determining yield, with some varieties being inherently higher yielding. This is why it's important to trial different varieties and take yield data. One variety might be a Dutch greenhouse diva, cranking tomatoes out when all its demands are met, but it might not be a good yielder in your rough-and-tumble hoophouse. Alternatively, there might be some scrappy varieties that will do well under some stress in a less-than-optimized greenhouse environment but go vegetative under optimal conditions. It is important to make sure that a variety is a good yielder under your particular conditions.

As is typical with most fruiting crops, the largest potential yields are with the largest-fruited beefsteak varieties. You can expect to get 80 to 90 percent of beefsteak yield from a cluster tomato, 40 to 60 percent from a cherry or cocktail tomato, and 25 to 50 percent from a grape tomato.

Harvest

Protected growers have a few harvest strategies that differ from those of field growing. One advantage of harvesting protected culture tomatoes is that they should be very clean, so they don't have to be washed. This enables harvesting with some of the green plant parts intact. Leaving calyxes or trusses attached to the fruit makes for nice presentation, and helps differentiate your produce from field tomatoes.

The most famous example of this is the TOV, aka cluster tomato. The idea behind growing TOVs was that the appealing-looking green stem would symbolize freshness to consumers. Indeed the TOV has gone from a nonexistent category of tomato a couple of decades ago to one of the largest categories of greenhouse tomato production today.

For tomatoes that are harvested loose, the calyx (the little green hat at the stem end of the fruit) can be left on. This is usually removed from field tomatoes so the stems won't puncture other tomatoes that are stacked around them. Protected culture tomatoes are usually picked into the same containers they will be sold in, to eliminate the extra labor of sorting them from bulk into sale packaging. Since loose tomatoes are often packed into single layer flats, the calyxes can be left on without danger of puncturing other tomatoes.

There has been a great deal of debate over what the "right" stage of ripeness is to pick tomatoes. When to harvest depends on what your goals are and what you want to do with the tomatoes.

The bad reputation that tomatoes picked green and ripened off the vine have acquired is deserved, and this has inspired a countertrend to do the opposite and pick tomatoes completely ripe. I believe that the best stage of ripeness for harvest lies somewhere in between the two extremes.

If you are going to pick a tomato and eat it right away, there is nothing wrong with picking it completely ripe. But there is a point after the ripening process has started, before it has achieved full color, when a tomato will achieve full flavor even off the plant. During the growth process and the beginning of the ripening process, a tomato receives a lot from the plant. But during the last days of ripening, it doesn't receive as much from the plant anymore and can finish ripening on its own.

It is usually in the interest of commercial growers to pick their tomatoes before they have reached full ripeness. Growers who need to ship their tomatoes often pick them less ripe depending on how far their tomatoes have to travel until they reach the end user. Some wholesalers specify slightly underripe tomatoes so they have some shelf life. Other reasons to pick tomatoes less than fully ripe include the fact that as they ripen, their skin loses its elasticity. When tomatoes are unripe, the skins are fairly elastic, which is what allows them to swell to their final size without cracking. But as the tomato begins to ripen and change color, the skin loses its ability to expand. Picking before full ripeness reduces last-minute cracking. Another reason to pick tomatoes before they are completely ripe is because adverse weather can actually hurt tomato flavor. You wouldn't want your tomatoes to be stored at very cold or hot temperatures off the vine. As long as fruit is in the greenhouse, it is vulnerable to damage from bad weather, splitting, and pests and diseases. So if you can anticipate excessively hot or cold temperatures, be proactive and get all the fruit that is close to being ready off the vines.

Determine Your Own Ripeness Preference

So what's a grower to do? Prioritize flavor or shelf life? Unfortunately, most of the things that improve handling reduce flavor, and vice versa. Fortunately, unless your supply chain is really long, you can have both flavor and shelf life. Part of the equation includes having a nice tomato storage area, a ripening room with good tomato storage and ripening conditions (see "Post-Harvest/

Flavor: An Opportunity

Sugars take extra energy to produce, so flavor tends to correlate inversely with yield. Because the largest producers tend to focus their efforts on yield and appearance, growing for flavor presents an opportunity not just for local growers, but for anyone who wants to carve out a niche for themselves as tasting better than the competition. Besides generating repeat customers, better flavor should command a higher price — because better-tasting produce is a higher-quality product, and because most of the strategies to improve flavor in tomatoes decrease yield.

Flavor is an especially contentious issue when it comes to tomatoes, because it's subjective. With other features of the plant, like yield, we have a hard number to point at. With flavor, not everyone can even agree on what tastes good. Flavor is a function of variety, moderated by the environment the tomato was grown in, as experienced by the person eating it. This means it's up to you to pick a variety that you think tastes good, or that you think your market will enjoy. Keep in mind that some of what eaters typically include in the realm of flavor are things like texture or mouthfeel. Some people don't like soft tomatoes; others don't like firm ones. Once you've picked a variety, there are a few things that you can do to maximize how good it will taste.

This is where disease resistance plays into the flavor discussion. You could be growing the best-tasting variety in the world, but if it doesn't have any resistance to the diseases you have, it may end up being unhealthy. And if it's unhealthy plant, it won't taste its best. Healthy leaves and a healthy root system are essential for production of both yield and flavor. One way to look at it is that the roots are the pipelines and the leaves are the solar panels powering the sugar factory. If either one isn't healthy, tomatoes are not going to produce up to their potential yield or flavor.

I always think of this when people discuss which variety is the best tasting. There are big differences and it's fun to talk about our favorites. But the flavor is really the product of the variety and its interaction with the environment. This is the French concept of *terroir*, or all the environmental factors that go into the character of food.

If your favorite variety is an heirloom with no disease resistances, which you plant in an environment with high disease pressure, you may not get a single tomato off the plant, let alone a good-tasting one. This is an extreme example but one that is relevant in less dramatic fashion to protected tomato growing.

I've seen some growers kill the flavor of good-tasting varieties. And I've seen others get varieties with only average potential to taste good. The simplest example I can think of is this: If you are growing a Brandywine (or whatever tomato you prefer) and it's covered with leaf mold, its diseased leaves aren't going to have much photosynthesis going on. A leaf-mold-resistant variety

Storage" on page 158). So regardless of the conditions in your greenhouse, the tomatoes have the benefit of ideal storage conditions to help them reach their peak of flavor.

Here's a way you can come up with your harvest standards: First of all, think of your supply chain. Do your tomatoes have to travel a long way and then sit on a store shelf before someone eats them? Or do they just have to go a couple of miles down the road to the local farmers market? Having a short supply chain is one factor that local growers should take advantage of for flavor that stands out from tomatoes that come from far away. One good way to determine the stage at which you should harvest your tomatoes is to harvest one that is just barely starting to get some ripe color on the blossom end, take a with a more modest flavor pedigree growing right next to it might well have better flavor, under those conditions, than a Brandywine. Since leaves are the photosynthetic factories for sugars and everything else in the plant, if they are diseased the fruit isn't going to taste like much.

This doesn't mean that we should pick varieties based solely on disease resistance. I still grow some heirlooms because I like them and I have a market for them. But it's worth keeping in mind that varieties will also taste better if we can keep them healthy, and that is as important a part of the equation as pedigree.

Influence of Type on Flavor

There are a few generalizations that can be made about the flavor tendencies of certain categories of tomatoes. Generally speaking, yellow, orange, and pink tomatoes tend to be lower in acid than many of the red ones. Because I like tomatoes that are high in acid and sweetness, I tend to find a lot of the yellow ones bland. However, if we don't have any yellow tomatoes on the stand, we always have market customers who let us know they miss them. They remind me it is a good principle not to rely solely on your own tastes if you want to stay in business growing food for others.

Another generalization is that smaller tomatoes tend to be sweeter than larger-fruited cultivars. It's almost like each fruit has a certain potential for sugar, and it can either stay small and concentrated or get large and diluted.

Brix

Tomato varieties are sometimes marketed as having a certain amount of brix. Brix is a way to measure how much sugar is in a solution, so when someone says their tomato has a high brix they are saying it is sweet. However, a stated level of brix is not a completely reliable indicator of how sweet a variety will be, because the level even in tomatoes of the same plant will vary over the course of the season depending on how healthy the plant is, light levels, growing conditions, and more. Brix should also not be confused with good flavor. A tomato could be very sweet and still have poor flavor.

Brix can be a good way to monitor the quality of your own produce throughout the season. Brix is measured by an instrument called a refractometer; they are cheap and easy to use. All you have to do is put some plant sap on the refractometer and look through the instrument to see what the brix is. Just make sure you get one made for plants, as there are refractometers made for measuring a variety of different fluids that have different calibrations. If you take readings on a regular basis and they start dropping, something is likely going wrong with your plants and sweetness is dropping.

picture of it, and put it wherever you ripen your tomatoes. A few days later, when the color has come halfway up the sides of your first tomato, pick another tomato fresh from the vine that matches the amount of color as the one you one you picked a few days ago. Label the tomatoes so you know which is which, since now you have two tomatoes at the same stage of ripeness. Wait a few more days until both tomatoes are fully ripe (or however you like to eat your tomatoes), and pick one fresh from the vine that matches the two tomatoes you already have. Now do your own taste test. Try all three tomatoes. It would be even more interesting to let other people taste-test them. Since they don't know that the tomatoes were harvested at different stages of ripeness, and they all look the same at this point, it would be a way to get a blind opinion of whether people can tell the difference between the tomatoes that were harvested at different stages.

My guess is that the tomato that was harvested with hardly any color is not going to taste as good as the tomato harvested with half color. But I bet the tomato harvested at half color tastes pretty similar to, if not the same as, the tomato that was harvested ripe from the vine. But maybe not; that's the beauty of tasting for yourself and developing your own flavor protocol. If you develop your own flavor standards, then you can communicate them to employees and help ensure that tomatoes will be picked to your standard.

Post-Harvest/Storage

Small-fruited tomatoes are usually picked loose into pint containers. Large-fruited and tomatoes on-the-vine are also usually picked into the container they will be sold in. This is most commonly a flat that holds a single layer of tomatoes. For large, soft-fruited varieties like heirlooms, it may help to put a thin layer of foam wrap (like bubble wrap with smaller bubbles) in the bottom of the flat to cushion the tomatoes from bruising under their own weight.

Tomatoes are most frequently packed with their blossom ends down, so the calyx is visible. Some growers prefer to pack them with the calyx down and blossom end up so they don't bruise as much. The blossom end is the softest part; since that is where ripening begins, it is a few days ahead of the rest of the tomato.

Exposure to temperatures below 59°F (15°C) will give tomatoes a mushy, mealy texture and damage the flavor compounds, dulling the flavor. Tomatoes give off ethylene, so don't store them with sensitive crops like lettuce and cucumbers. Since tomatoes have a higher storage temperature than most other crops, and don't store well with a lot of other crops, it's a good idea to have a ripening room where freshly picked tomatoes can be held until they are sold. Some growers cool this room for maximum shelf life, whereas others just try to keep it from getting too hot. Anywhere that is out of direct sunlight and not too hot will be better than nothing. Tomatoes will continue to ripen more quickly when kept at warmer temperatures.

On that note, remind your customers not to put their tomatoes in the refrigerator. It doesn't matter how fantastic a tomato you have grown if they start destroying the flavor as soon as they get it home. No wonder some people don't like tomatoes – their parents probably put them in the fridge. Customers enjoy learning these kinds of tips if they don't already know them, and you may just preserve your tomato reputation.

CHAPTER TEN

Cucumbers

Cucumbers are a very fast-growing crop. When grown quickly at high temperatures, they can take as little as twenty-one days to go from seed to transplant stage, and another twenty-one days to go from transplant to first harvest. Under hot conditions, mature plants can grow 4 inches (10 cm) or more in a day. As long as they are kept well watered, cucumbers plants can tolerate hotter temperatures than tomatoes.

The fast growth rate means cucumbers may require more frequent attention than the other vining/fruiting crops. During periods of fast growth, cukes may need pruning or trellising more than once a week. Small-fruited varieties like cocktail cucumbers may need to be picked once a day or more to keep them from getting too big.

Varieties

There are two specific adaptations that are unusual in field cucumbers, which help maximize the potential of cukes grown in protected culture. All cucumbers bred for protected culture have these adaptations.

Gynoecious

The term *gynoecious* means that all the flowers produced by the plant are female. This is in contrast with most cucurbits, which have male and female flowers on the same plant (monoecious). Only female flowers result in fruit; the males are just for pollen.

It is particularly important for protected culture cucumbers to be gynoecious, since the plants are usually pruned to remove the majority of the flowers. It is crucial to know that every remaining flower has the potential to turn into a fruit. Pruning is a way to control the number of flowers that are set on the plant. Most pruning schemes are designed to leave a particular number of fruits to create a balanced plant. There would be no way to prune accurately if you had no idea how many of the flowers on the plant would turn into fruit.

Another important aspect of gynoecious cucumbers is that they tend to yield more per plant than varieties that have male and female flowers, since every single flower can make a fruit. I know from experience that growing monoecious varieties in the greenhouse can lead to very low yields. Over my years of doing variety trials, we would occasionally throw a monoecious variety into greenhouse trials, and the yields were always low. The only time growing a monoecious variety in protected culture should be considered is if the plants will be allowed to grow sprawled and unpruned in a structure that either has bees inside or is open to pollinators from the outside.

Parthenocarpic

Parthenocarpy describes the ability of a cucumber plant to set fruit without pollination. Most cucurbits do not set fruit unless pollination has occurred. This is especially important in structures growing gynoecious varieties as described above. Since pollen comes from the male flower, a greenhouse full of all-female cucumber plants

Ideal Cucumber Climate

OPTIMAL TWENTY-FOUR-HOUR AVERAGE TEMPERATURE

A good twenty-four-hour average temperature for cucumbers is 70°F (21°C).

MIN/MAX TEMPERATURE

Temperatures below 61 to 63°F (16–17°C) will shut the plants down. Long European types are especially sensitive to cold. Cukes can take some prolonged hot weather, but it's best to avoid temperatures above 86°F (30°C). Hotter than this and transpiration rates are so high it can be difficult to keep cucumbers watered enough to prevent wilting. It is not unusual to notice a little wilting on very hot days, but it's important in these cases to get the plants some more water. If plants wilt for a prolonged period, individual leaves may develop dead areas where the tissue became too damaged to recover.

84
82
80
78
76
74
72
70
Temperature
Day
Night
Time of Day

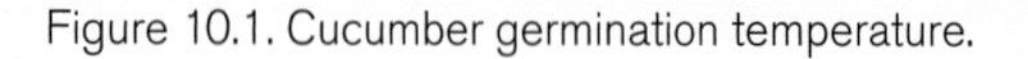

Figure 10.1. Cucumber germination temperature.

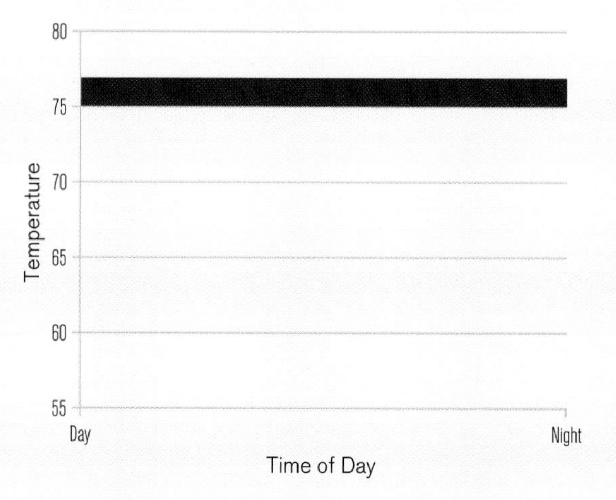

Figure 10.2. Temperature for growing cucumber seedlings the first week after germination.

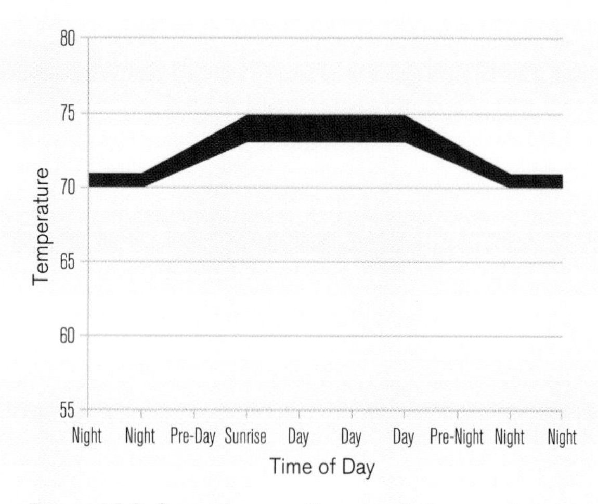

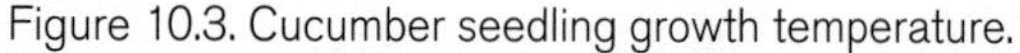

Figure 10.3. Cucumber seedling growth temperature.

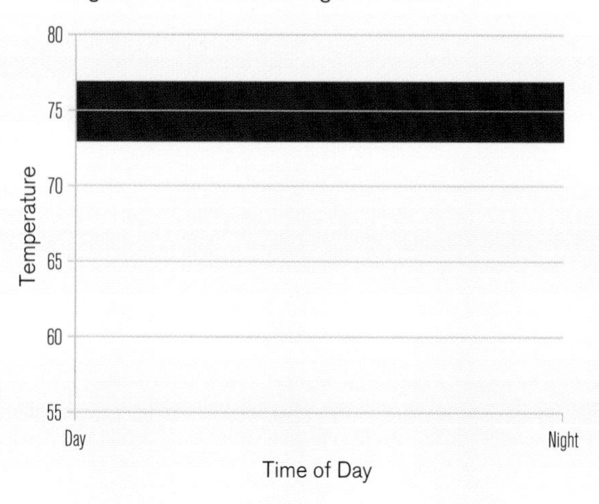

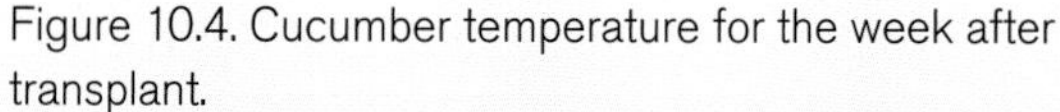

Figure 10.4. Cucumber temperature for the week after transplant.

CLIMATE

For the first week after transplanting to the production greenhouse, use a flat temperature profile of 73 to 77°F (23–25°C) day and night. After the first week, maintain the same daytime temperature but lower the night temperature to between 70 and 73°F (21–23°C).

When cucumber harvest starts, maintain daytime temperatures of 70 to 75°F (21–24°C), and night temperatures of 63 to 68°F (17–20°C). When the plants are fully loaded, try to maintain a temperature of 75 to 80°F (24–27°C) during the day and 63 to 70°F (17–21°C) at night.

Cucumbers are happiest when the root zone is in a range from 65 to 68°F (18–20°C). If temperatures go far outside this range, growth may be adversely affected.

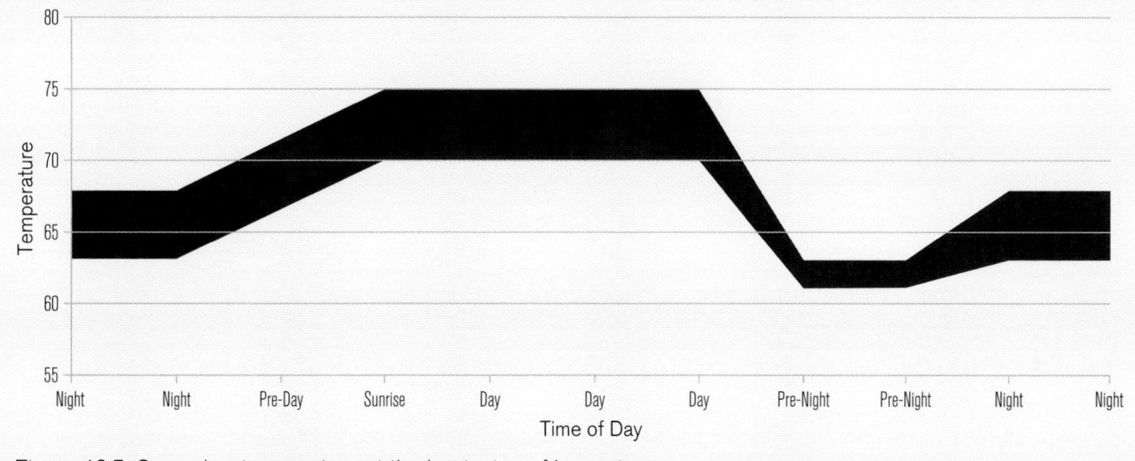

Figure 10.5. Cucumber temperature at the beginning of harvest.

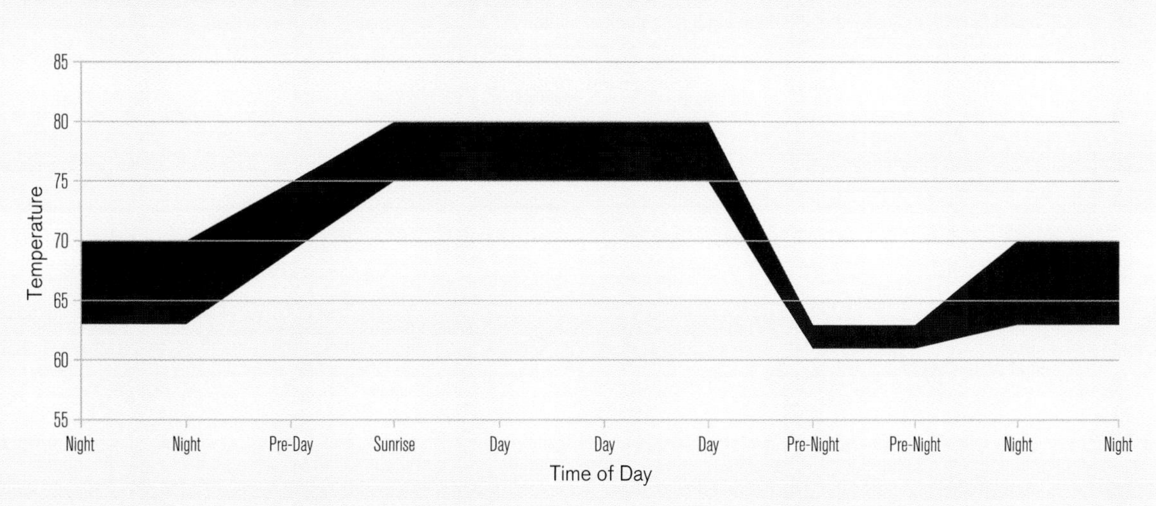

Figure 10.6. Cucumber temperature at fully loaded harvest.

that aren't parthenocarpic wouldn't set any fruit – none of it would be pollinated.

The other advantage of growing parthenocarpic varieties is that they have the potential to be seedless. I say "the potential to be seedless" because even though they will set fruit without pollination, they will still make seeds if they get pollinated. This could occur if there were varieties with male flowers planted in the same greenhouse or just outside the greenhouse. Pollinators could move the pollen around inside the greenhouse, or bring it in from the outside.

Many growers really want to prevent their cukes from being pollinated. Preventing pollination is yet another reason not to grow monoecious varieties in protected culture. Sometimes if a fruit is only partially pollinated, it will be deformed, making it unsalable. The part of the fruit with seeds may swell and grow faster than the part without. The other disadvantage of having parthenocarpic cucumbers pollinated is that with the development of seeds, you give up the advantage of being seedless.

If you harvest any type of cucumber with hard seed, that fruit is overmature and past its peak flavor. But even immature seeds diminish cucumber flavor. They also diminish the texture of the cucumber, since the area around the seeds expands more and develops a softer consistency than if a cucumber remains unpollinated. Parthenocarpic cucumbers that are unpollinated have smaller seed cavities that stay crunchier than cucumbers with developing seeds. This is one of the important flavor advantages of protected culture cucumbers.

Long European Cucumbers

The overwhelming majority of cucumber varieties bred specifically for greenhouse production are of the long European type, otherwise known as English or Dutch cucumbers. There are probably more long European varieties developed for protected production than all the other greenhouse types combined.

European cucumbers are the 10- to 16-inch (25–40 cm) long cucumbers that are normally shrink-wrapped when sold in the grocery store. They are spineless and have small longitudinal ridges along the fruit. European cucumbers became popular with consumers due to their thin skin and lack of seeds. There is very little prep time and food waste since seeds and skins do not need to be removed. They can be eaten whole, and flavor tends to be very good.

European cucumbers are difficult to produce in the field, since rocks and insects can scar the thin skin and they are so long they tend to bend and curl when in contact with the ground. Due to their thin skin they need to be shrink-wrapped to prevent dehydration over long-distance shipping, though this may not be necessary for local selling. They became the preferred greenhouse type because of their flavor, and they gave protected growers something to set their produce apart from field cucumbers.

Cocktail/Mini Cucumbers

The cocktail cucumber is the smallest type. They are typically around 4 inches (10 cm) long, and are popular because they are very crunchy, thin-skinned, and easy to eat out of hand. They are usually sold pre-packaged, either bagged or shrink-wrapped to a cardboard tray. Flavor tends to be excellent, and this type has become very popular with the trend of snack vegetables.

In between the size of cocktail cucumbers and long Europeans are many varieties that are broadly known as "mini" cucumbers. Most of the mini types are the result of crosses between long Europeans and smaller types. The result has many of the characteristics of a European cucumber in a smaller package: thin skin, good flavor, lack of seeds or spines, and longitudinal ridges. There are a lot of mini varieties that can be harvested anywhere from slightly larger than cocktail cucumbers up to 8 or 9 inches (20–23 cm) long. This should give growers a lot of flexibility – regardless of the fruit size their market prefers – to provide a delicious, thin-skinned seedless cucumber in whatever package their market will accept.

The variety Socrates, which I grow on my own farm, is a good example of this. Socrates is about as large a mini cucumber as I have seen, topping out at about 8 inches (20 cm) long before it becomes bloated. This is larger than the breeder ever intended for it to be harvested but it works out well for us because at that size it looks like

Figure 10.7. My son enjoys a cocktail cucumber fresh from the vine in one of our hoophouses. Note the white ground cover. Among the factors driving the current trend in snack vegetables are their good flavor and ease of eating out of hand, which make them popular with kids and parents alike.

Figure 10.8. A pickling cucumber being grown by the high-wire (single-leader) method in the author's hoophouse.

an American slicer without the spines and thick skin. A lot of people are creatures of habit when it comes to food. Many people who are used to eating thicker-skinned cucumbers will continue to buy them regardless of the fact that they are not the best tasting. But since an overgrown mini like Socrates looks so much like a slicer, we are able to sell many of our customers a Socrates instead of the American slicers they are used to.

Though we don't make any more money off a Socrates than an American slicer, I want to provide my customers with the best possible eating experience. I do hope they will buy more if the cucumbers are really delicious, but even if they don't it makes me happy to give my customers something great to eat. Socrates is a way for me to give customers who wouldn't otherwise buy a long European cucumber a higher-quality product in a package they will accept.

Pickling Cucumbers

A number of pickling cucumbers have been developed over the years for protected growing. Most of them are of the European type, probably because most greenhouse breeding is done in Europe. European pickling cucumbers have more, smaller spines than the pickling cucumbers most Americans are used to. Both types tend to be 5 to 6 inches (13–15 cm) long, depending on when they are picked.

There are now some American-type pickling cucumbers available for greenhouse growing. There isn't much difference in flavor between the two types, and most pickling cukes' flavor is quite good for fresh eating.

American Slicers

The American slicer used to be rare in protected growing, but not anymore. There's only a fraction as many varieties of slicers compared with long European types, but there are more than ever available now. I think breeding companies used to shy away from American slicers because their flavor isn't the best, and because they thought that greenhouse growers wanted to grow something that would set their produce apart from field cucumbers. But with the expansion of greenhouse growing in the United States, some protected growers have demanded slicers. Flavor in this category isn't a huge improvement over what is grown in the field, though some of the greenhouse slicers do have thinner skins, which will help.

Just like the field version, American slicers for the greenhouse tend to be 7 to 9 inches (18–23 cm) long and dark green, with thick skin and few spines. One advantage to growing American slicers in a greenhouse is that you can choose a relatively thin-skinned variety, since thick skin isn't necessary to protect the fruit from pests and being on the ground as it is in the field.

Winter Versus Summer Varieties

Especially in northern areas, different varieties of cucumber are grown in the winter than during the rest of the year. Summer varieties have the highest level of powdery mildew (PM) tolerance possible since disease pressure is highest in the summer. Currently, varieties with the highest level of PM tolerance have a tendency to develop necrotic spots on the leaves under winter conditions. Since pressure isn't as high during the winter, growers use varieties with lower PM resistance that have been bred to do better during times of the year with lower light.

When to use winter varieties varies greatly by region. In warmer areas, powdery mildew pressure and light levels may stay higher for a greater portion of the year, necessitating longer use of summer varieties. At some point in the future, breeding may improve to where there are highly PM-resistant varieties for the winter. Be aware of these varietal distinctions and ask your seed supplier which will do better in the winter in your area.

Single Versus Multifruited

Some varieties are bred to produce a single flower per node; others are bred for multiple flowers per node. For larger-fruited varieties (7 inches/18 cm and larger), it's desirable to form only one fruit per node. This is because the plant can only support one large fruit per node, so multiples would have to be pruned off. Being single-fruited is a labor-saving adaptation in larger-fruited cucumbers. Even varieties that are mostly single-fruited will sometimes produce multiple fruits, so it's still important to tell workers to remove additional fruits. For smaller-fruited varieties, it's desirable to be multifruited. The fruits are so small that the plant can support multiple fruits per node. Being multifruited can help improve the yields of smaller-fruited types.

Cucumbers from Seed to Sale

Once you've chosen your varieties, it's time to think about how to make the most of them. Below are cucumber-specific recommendations to take you from seed to sale.

Crop Timing/Cycles

One of the advantages of high-wire cucumber production is that it leverages the benefits of a long-season cropping schedule, namely an extremely long harvest season, from just two establishment periods per year. When the crop is kept healthy, the grower is allowed some flexibility, since the cropping period can be extended if the market is good. For growers who do not plan on cropping year-round, as with a long-season tomato crop, this may mean that if prices (or weather in the case of hoophouse

Figure 10.9. A snack cucumber being pruned to a single leader. Note that there are at least two flowers per node, which will increase yield on small-fruited varieties like these.

growers) are good they can continue heating and keeping the crop alive if prices justify it.

Umbrella cukes, on the other hand, are usually grown in three to four crops a year, each cycle three to four months in length. Because the plant only reaches a certain size, yield cannot be extended indefinitely the way it can in high-wire cukes.

Propagation

Because cucumber plants grow so quickly, some growers skip the step of planting into plugs and seed directly into larger blocks. If cucumbers are sown into plugs, they usually need to be transplanted into blocks within a week of germination. Cucumbers resent having their roots disturbed, so do not delay this timing too much and allow them to get root-bound.

Germinate cucumber seeds at 82°F (28°C). Upon emergence, grow them for the first week at 75 to 77°F (24–25°C) both day and night. After the first week, when they go into blocks or larger pots (if they didn't start there), run a day temperature of 73 to 77°F (23–25°C) and 70 to 73°F (21–23°C) at night.

Grafting

It is possible to graft cucumbers, including to other members of the cucurbit family. There are some interspecific pumpkin/gourd hybrids that are used for grafting cucumbers just as interspecific hybrids are used for tomato rootstocks. With currently available varieties, the benefits are limited to soilborne disease resistance, without the yield boost commonly seen in grafted tomatoes.

At this point, cucumber rootstock options are not as developed as they are for tomatoes, and there is little commercial cucumber grafting in North America. In Europe and especially Asia, however, grafting cucumbers is a common practice. It is so much more prevalent there because many of their vegetable production areas are contaminated with soil diseases that make it difficult to produce cucumbers on their own roots. This is also the case in the watermelon industry in the United States, where grafting is common in some of the main production areas due to disease infestation. So growers in these regions are forced to graft if they want a crop at all. As cucumber rootstocks progress to the point where tomato rootstocks are – adding both disease resistance *and* a yield boost – I think it is likely cucumber grafting will become the norm as it has for tomatoes. Until then, cucumber grafting in North America will probably remain confined to growers who have to do it for disease resistance.

When considering cucumber grafting, or cucurbit grafting in general, it is very important to have tried the rootstock/top variety combination on a small scale before going into production. Sometimes cucurbit rootstocks will change the flavor or other characteristics of the top variety. This has been a problem in melon grafting, where some rootstocks give melons cucumber or other off flavors.

Transplanting

Cucumber seedlings are ready to transplant into the production greenhouse when they have anywhere from three to five true leaves, which may be twenty-one or more days after germination, depending on how much heat is used. To keep larger plants in propagation for as long as possible, make sure they are spaced so they don't shade one another. Otherwise they will begin stretching. Stake seedlings if necessary to keep them from flopping over.

Spacing

Umbrella-style cucumbers can be planted anywhere from 1.2 to 1.8 plants/m^2. A number of different crop layouts can be used to achieve this spacing, with wider pathways using tighter in-row spacing and vice versa. A starting point would be to use our standard greenhouse setup with 2-foot-wide (60 cm) rows and 3-foot-wide (90 cm) pathways, with plants 18 inches (45 cm) apart in a single row. Every other plant goes to each side of the double row, so the heads are 3 feet apart along each wire. The same plant spacing can be achieved with two rows, with plants 3 feet apart in each row. Increase or decrease in-row spacing to change density if desired depending on the season. For example, give the plants more space during the darker parts of the year, or crowd them when there is more light.

Usually, high-wire cucumber production shoots for about half the density of tomato production, with 1.8 plants/m^2. They can be planted in a single row, 1 foot (30 cm) apart with one head going to each side, for a final distance of stems 2 feet (60 cm) apart along the wire. Or they can be planted in double rows, 2 feet apart, with each row going to the corresponding wire. Reduce density if you expect low light levels.

If plants are to be grown when good light and weather are expected, you can increase density up to a level usually used for tomatoes, around 3.6 plants/m^2. Since cucumber plants are not usually double-headed, you could achieve this spacing with a single row of cucumber plants 6 inches (15 cm) apart, with every other plant going to an alternating wire. Or you could plant two rows of plants 1 foot (30 cm) apart, with each row going to the corresponding wire. I have grown them at this density in a hoophouse during the summer with good results.

Pollination

Pollination is not needed for cucumbers as long as you use parthenocarpic varieties. See "Varieties" on page 159, for more information.

Trellising

There are two main styles of trellising cucumbers: umbrella and high-wire. Umbrella cultivation has long been the main way of growing greenhouse cucumbers, but high-wire is gaining popularity. The trellis structure looks very much the same for either style; it's the height of the wire and how the grower manages that plant that differentiates the two styles.

Of course, you can grow cucumbers in a hoophouse letting them sprawl on the ground as they would in the field. This would require a minimal amount of labor to maintain, and is a viable option for hoophouses that don't have – or can't accommodate – trellising. Though letting cucumbers sprawl saves labor on trellising, cucumbers on the ground are much more laborious to pick.

Growing the plants up a trellis takes advantage of the vertical space in your structure. Making use of all that vertical space means the plants aren't competing with one another as they would if sprawled on the ground. High-density planting is an important protected culture technique for increasing yield.

High-Wire Versus Umbrella Production

Umbrella has been the standard way of producing cucumbers in the greenhouse industry for a number of years. It minimizes labor by turning the cucumber plant from a vine into a plant of determined size, by topping it when it reaches the wire, and harvesting most of the cukes off suckers (see "Pruning for Umbrella Trellising" on page 172).

One of the reasons high-wire production is gaining in popularity is because it works better in the taller types of

Figure 10.10. A young crop of long European greenhouse cucumbers grown umbrella-style. The plants have just reached the top wire, which is 6 to 7 feet (1.8–2.1 m) high, and are soon to be topped to let the laterals develop.

greenhouses that are being built now to take advantage of the efficiency of a larger air mass. I have grown high-wire cucumbers in hoophouses with a top wire at 9 feet (2.7 m), and I wouldn't want to go much lower than that. If your structure is any shorter, you would have to lower the plants before harvesting the cucumbers, putting the fruit in contact with the ground. Umbrella-style growing was developed for lower structures, so it's probably the best option if you have a low greenhouse. Umbrella can work in greenhouses with top wires as low as 6 to 7 feet (1.8–2.1 m).

The main benefit of umbrella-style cucumber production is labor savings. Because the crop is only grown and pruned to a certain height, the labor required early in the crop is mostly pruning; later, it's mostly picking. In high-wire, you are picking and trellising at the same time for almost the entire life of the crop.

The drawback of umbrella production is the amount of time out of production that must be spent establishing new crops, and the cost to clean out and replant greenhouses between crops. Between seed, propagation, medium, and labor to take out the old crop and put in the new, it's a significant expense to establish a new crop three or four times a year. These establishment costs are part of the reason why high-wire production is becoming more popular.

Another advantage of high-wire production is a higher overall yield. Everything else being equal, two high-wire crops per year tend to yield more than three or four umbrella crops per year. The cucumbers from high-wire production are also of higher average quality than umbrella cukes.

In high-wire production, all the cucumbers are harvested from the main stem of the plant, and those (known as stem fruits) tend to be longer and darker green than fruits that develop on the laterals. This may be more of an advantage for larger growers who are competing on the wholesale market. At times when the market is flooded and buyers can pick and choose, they will use quality to choose from among suppliers.

Regardless of the size of the grower, both methods have their advantages and can be profitable. The higher quality and yield of high-wire-style growing makes it competitive with umbrella production. Growers should choose based on whether their structure is tall enough to accommodate high-wire production, how much labor they want to put into the crop, and personal preference for one style or the other.

Adaptations for High-Wire Versus Umbrella

By and large, the same varieties can be used for umbrella-style cucumber production as for high-wire. As high-wire production becomes more popular, there are some varieties that are being billed as specifically for it. One of the main differences is selectivity. If a variety is described as selective, it means it's more likely to abort fruits when under a high fruit load. The plant chooses whether or not to support a particular fruit. This can be good for umbrella production, because cucumbers are the type of plant that will put too much energy into fruit at the expense of the plant when there is too much fruit set. Since fruit set is left more to the plant in umbrella-style growing, it may be a good thing for the plant to abort some fruits if it has a lot of blossoms and is in danger of becoming overloaded.

On the other hand, high-wire cucumbers are pruned much more than umbrella-style cukes. High and even production in high-wire is dependent upon every flower left on the plant turning into a fruit. Varieties bred specifically for high-wire production are not selective, so the grower can make the decision about which fruits to get rid of and which to keep, not the plant.

Early maturity is more important in varieties for umbrella production, since with multiple crop cycles per year, days to maturity multiply over the course of the year. For example, if you plant a variety that matures two days later, this adds up to eight additional days of the year that you'll be out of production than if you'd used an earlier variety.

Labor

The decision of which trellising method to use is frequently based on labor. Umbrella-style cucumber production requires less labor, in the range of a hundred hours per acre per week. This is because the early labor is mostly focused on trellising, and after the plants are topped the labor switches mostly to picking.

High-wire cucumber production requires labor similar that of tomato production: in the 140-hours/acre/week range. This is because pruning, lowering, and harvesting are all going on at the same time with this method. The plants also need to be de-leafed as they are lowered, which isn't a concern in umbrella cultivation. The propagation costs for high-wire production are lower, however, since two crops are grown per year instead of three or four. Another difference is that umbrella production tends to be more uneven, known as flushy, with fluctuations from week to week in the amount harvested, whereas high-wire production is more consistent.

Given these differences, you have to decide whether you want the higher labor, higher yield, and higher quality of high-wire production, or the lower labor, lower yield, and shorter crop cycles of umbrella-style production.

Pruning

When left to their normal growth habit, cucumber plants tend to set more fruit than they can support. This leads to flushy production, where many cucumber fruits are set and then a number of flowers are aborted in a row. One of the goals of pruning greenhouse cucumbers is to achieve a more consistent yield by limiting the number of fruits the plant is feeding at any one time. This can lead to higher overall yields, as

Figure 10.11. A typical node on a cucumber plant. You can see how there is one female flower with a baby cucumber at the base, one sucker, and one tendril coming toward the camera. The leaf is off to the left.

Figure 10.12. The same shot of a typical cucumber node panned out so you can see the leaf and more of the tendril.

Figure 10.13. An example of suckers that have gotten too big. There are big suckers growing out from the two nodes in the middle of the picture. These should have been removed when they were 1 or 2 inches (2.5–5 cm) long.

a balanced plant may stay vigorous for longer than a plant that becomes overloaded.

The typical cucumber growth habit is to produce one tendril, one sucker, and one or more flowers at the node where each leaf is attached. Different pruning regimes are used for each production style.

Fruit Pruning

Regardless of which trellis method you use, when growing large-fruited cultivars it is standard to remove the first few flowers from each plant before they develop into cucumbers. Cucumber plants may start producing flowers at a very small size, before they can support a fruit load. If a small cucumber plant develops a high fruit load, it may set the plant back permanently. Removing a few flowers in the beginning to let the plant get bigger before trying to set fruit leads to a higher overall yield.

The size of the cucumber determines how many fruits need to be removed in the beginning, and how many should be allowed to set per node. Large-fruited cultivars, like long European types and 7- to 8-inch (18–20 cm) American slicers, should not be allowed to set fruit on the first 2 feet (60 cm) of the vine.

Some growers remove even more flowers, in winter production cycles or if the plants aren't particularly strong, for example. It's a trade-off, though, because often the first cukes of the season may bring the best prices. Whether to prioritize early fruit production or rapid establishment is a judgment call by the grower.

Regardless of how high you allow them to begin forming on the vine, large-fruited cultivars should only be allowed to develop one fruit per node. Most large-fruited cukes are bred to only produce one flower per node anyway, but if more than one fruit per node does develop, it can lead to too high a fruit load on the plant.

Varieties smaller than the 7- to 8-inch (18–20 cm) American slicer size can be allowed to set fruit freely from the beginning. These smaller-sized fruits are not as much of a burden on the plant, so they don't need to be thinned aggressively unless the plants seem weak and need a break at the beginning. Multiple fruits per node can be allowed to develop in these smaller cultivars. Being multifruited is desirable for small-fruited varieties because getting more than one fruit per node will increase yield in these types.

Pruning for High-Wire Trellising

High-wire cucumber production is very similar to lowered and leaned tomato production. This may appeal to some tomato growers because they can manage their cukes in much the same way as their tomatoes. The goal is to produce a single long vine that can be lowered as needed to keep it below the wire. Each plant gets a piece of twine attached to the overhead wire with a tomahook or spool that can be lowered and leaned. Vines are twisted around the twine or clipped to it as they grow upward.

Figure 10.14. What the head of a greenhouse cucumber plant looks like, with smaller, immature leaves surrounding the main growing point at the top of the plant.

Pruning for high-wire cucumber production is simple. The goal is to limit each plant to a single stem, with one flower per node. During times when the plants are under stress or too generative, you may remove every other flower to allow the plant to direct more energy toward vegetative growth. At the beginning, remove the fruit from the bottom 2 feet (60 cm) for large-fruited varieties (as recommended in chapter 6).

The pruning regime is very simple and stays more or less the same for the life of the plant. At every node, the sucker and tendril are removed, and one or more flowers are left to develop into cucumbers based on the size of the fruit (see "Fruit Pruning" on page 170). It's easy to break the sucker and tendril off cleanly with your fingers when they are in the 1- to 2-inch (2.5–5 cm) range.

Never remove all the suckers from the plant at once. Always leave a few at the head of the plant. If the head gets broken off or damaged by a pest, you can use one of the suckers to develop a new head.

Tendril Pruning

It may help to remove the tendrils, especially in high-wire production. It's difficult to lower and lean the plants when their tendrils grab one another. It's also a problem when a tendril grabs an immature fruit. Fruits that are grabbed by tendrils are usually unmarketable since the tendril constricts their growth and causes scarring.

It's not necessary to remove the entire tendril. Snapping off at least half will make it unable to grab. This can be done with one hand when suckers are being removed. Tendrils are not very strong; you can grab the base with thumb and forefinger, with the end between your palm and pinkie and ring finger, and snap it off quickly that way.

Figure 10.15. A cucumber plant being grown high-wire-style in the author's hoophouse. The plant will be kept to this single-leader style of pruning for the entire season. This is an example of good fruit set, with one large cucumber per node.

Figure 10.16. A tendril grabbing a cucumber. This will probably scar and deform the fruit.

Pruning for Umbrella Trellising

Pruning and trellising for umbrella-style production is a little more complicated, but it works much better in lower structures with wires in the 6- to 7-foot (1.8–2.1 m) range. Workers can typically maintain plants of this height from the ground. One piece of twine per plant is tied to the overhead wire. The vine is grown up the twine by twisting or clipping, removing the suckers on a regular basis as the vine grows up.

When to allow cucumbers to start developing on umbrella plants depends on the time of year. After clearing large-fruited cukes off the initial 2 feet (60 cm) of stem, many growers let a cuke develop at every other node up to the wire in sunny parts of the year. For winter growing in low-light areas, the plant may be limited to as few as three cucumbers on the main stem below the wire in the darkest time of the year, with progressively more cucumbers on the stem for later plantings.

As the plants approach the top wire, make sure to leave the three suckers immediately below the wire. Let these begin to grow out, and as the head of the plant grows over the wire, cut it off just above the wire. Topping the plant will push the energy that was going to the head to the suckers left on the plant, also called laterals.

These first laterals to grow out are called the primary laterals. When the primary laterals are long enough, guide them over the wire so the wire supports them, instead of them pulling down on the main stem. Going over the wire will help support the weight of the fruit that develops on the laterals.

At this point, you're almost done pruning. Do not prune the suckers from the primary laterals. These will become the secondary laterals, where most of the fruit is set. When the primary laterals grow down to about chest height, pinch the tips off to stop their growth and you're done pruning.

Secondary laterals and cucumbers will continue forming on the primary laterals until production slows down two to three months after it starts. Umbrella trellising essentially turns the cucumber vine into a determinate by limiting its upward growth. The plant will only remain productive for so long after being topped. Some growers decide to start seed for the next crop when they notice yield starting to slow down. Other growers simply start plants based on a pre-planned cropping schedule that they know works well with their seasons.

Figure 10.17. A long European cucumber plant in umbrella production shortly after being topped. You can see where the main stem was cut just above the leaf above the wire. You can also see the laterals that are being left to develop to the left and right of the main stem. When they're long enough, they will be thrown over the top wire to form the "umbrella."

Flavor

Cucumber flavor isn't hugely variable and thus isn't as a hot topic as, say, tomato flavor, though there definitely are better- and worse-tasting varieties.

Most greenhouse cucumber varieties taste pretty good. They've been bred not to have the things that people don't like about cucumbers. Some people have trouble digesting the seeds and thick skins of cucumbers. So you are at an advantage when it comes to flavor with protected culture cucumbers because your cukes should be seedless, and most have been bred for thin skin.

Thicker skins are typically a feature of field cucumbers, because they offer more protection from damage by pests and even from lying on the ground, which can cause marks and deformation on thin-skinned fruits. Though variety selection is important for finding a good-tasting greenhouse cuke, there isn't a lot of variation among them. Just by growing greenhouse varieties you are ahead of the game, which should be enough to keep customers coming back.

Pests, Diseases, and Disorders

Cucumbers are affected by many of the same pests and diseases as the other greenhouse crops, plus a few unique afflictions (see appendix B).

Cucumber Beetles

There are striped and spotted species of cucumber beetle, both of which are major pests. Some regions have both species; some are home to one or the other. Both beetles are yellow with black markings, and both can be a real pain for protected cucumber growers because they hatch out early in the spring, when there aren't many cucurbits in the field for them to eat. Early protected culture cukes may be the only thing on the menu, drawing all the beetles in the area.

In high enough numbers, they can cause significant damage to young plants by chewing the foliage, and they can be really damaging to young and old plants alike by transmitting bacterial wilt from plant to plant. Bacterial wilt is a disease that can cause healthy plants to suddenly collapse overnight. If you go into your house and see random plants that have completely wilted that were fine the day before, it's probably bacterial wilt. Remove affected plants as quickly as possible since they will not recover and may serve as a source of the disease if beetles continue to feed on them and move to other plants.

Another significant problem is caused when cucumber beetles chew on the fruit. Even if they don't eat much, their chewing will cause scar tissue that can mar the fruit. As the fruit grows, so does the scar, which may make a cuke unmarketable. The scar tissue is also occasionally less flexible than the rest of the cucumber, which may cause fruit to curve.

Because they can be difficult to manage once they get in, excluding cucumber beetles with screens is a good way to deal with them. There are also some organic and other sprays that can reduce their numbers, but beetles are difficult to get rid of.

Thrips and Whiteflies

See the basics on these pests appendix B. Because of their very small size, they are harder to exclude with screens than cucumber beetles without reducing ventilation drastically. When they get out of hand, their chewing and honeydew make a mess, and they can transmit diseases from plant to plant. There are a lot of beneficial insects that are effective against these pests.

Bacterial Wilt

Bacterial wilt is a disease that can cause plants to wilt suddenly and irrecoverably. See "Cucumber Beetles," above, their most important vector. There is no treatment, other than removing affected plants to minimize the spread by workers or insects. The disease is more common in some areas than others.

Downy and Powdery Mildew

Both downy and powdery mildews can be a problem in cucumber crops. Use resistant varieties as much as possible. Disease pressure varies seasonally.

Crooked Fruits

Crooked fruits, otherwise known as crooking, are a problem because most markets want straight cucumbers. The longer a cucumber is, the more prone it will be to this problem. Crooking may be caused by physical interference with fruit development or mechanical damage, such as a fruit being grabbed by a tendril or chewed by an insect. In the latter case, the cucumber may curl

Figure 10.18. Cold injury caused by water evaporating off the cucumber when the sun hits it in the morning. This can be hard to control in hoophouses, but is preventable in greenhouses by not allowing condensation to form, and by gradually warming temperatures in the morning instead of allowing them to spike when the sun comes up.

around the scar tissue that forms where there has been mechanical damage.

There can also be environmental factors, such as wide fluctuations in the amount of moisture in the root zone or air, or rapid fluctuations in temperature. To reduce the number of environmental causes, keep temperature and watering as even as possible.

Fruit Burn

If you see scarring down the length of a cucumber that doesn't appear to be due to insect or other mechanical damage, it could be due to evaporation. Cucumbers are very sensitive to cold temperatures. If condensation causes moisture to form on the fruits overnight, evaporation of the water in the morning can burn the cucumber fruit, causing scarring. Just as evaporating sweat cools us off, evaporating condensation can cool a cucumber down to the point where it damages the skin of the fruit. This tends to happen on the east side of rows, as this is the side that gets hit directly by the rising sun. Prevent fruit burn by eliminating condensation with heating, and gradually warming the greenhouse up so it's at daytime temperature by the time the sun comes up (see chapter 3).

Burnt Heads

Burnt heads is a disorder specific to pruned, trellised cucumbers. It can result from a wide variety of stresses that interrupt the flow of nutrients to the head of the plant, including temperature (too hot or cold), watering (too wet or too dry), humidity (too windy or humidity too low), growth that is too rapid due to excess fertility, or a heavy fruit load. It can be frustrating to diagnose a disorder with so many causes. However, burnt heads can usually be traced back to one or more of these stressful conditions, the effects of which build on each other. It has the same root cause as blossom end rot in tomatoes: not enough calcium reaching the head of the plant. The damage produced by this condition gives the newest leaves and tendrils a dry, curled, "burned" appearance.

Another possible cause of symptoms that resemble burnt heads is mite damage. Some species prefer newly emerged leaves and can cause a disproportionate amount of damage to the head of the plant. It's important to scout and make sure burnt heads are not due to mites, because they can easily get out of control if they go undetected.

Harvest

Frequency of harvest in cucumbers is determined both by variety and by how fast the plants are growing. The goal of cucumber harvest is to pick the cucumbers once they have reached a minimum size and degree of maturity, before they have become overmature. There is some latitude in when to pick cucumbers, since they are picked when *horticulturally* mature, not botanically mature (in other words, ripe).

Consider the difference between picking a ripe tomato or pepper (skin has changed color, seeds are mature) and picking a cucumber (based on size, not ripeness; seeds should not be mature if present at all). In greenhouse cukes, the matter can be confused, since they are parthenocarpic and gynoecious and don't set seed. But we can still say that overmaturity corresponds to the stage of maturity in a cucumber when it would otherwise have hard seed. This definition still allows for some latitude in when to harvest. A particular variety may be harvested slightly larger or smaller, once it has reached horticultural maturity, based on grower preference, buyer preference, and stage of growth. Cucumber plants in a more generative stage will tend to produce longer, larger fruits.

Overmaturity in cucumbers corresponds to bloating. This is because cucumbers grow to a certain length before doing most of their diameter expansion. So an overmature cucumber is one that has reached its final length and increased too much in diameter. You want to pick cucumbers at the desired length, before they have begun bloating, because flavor will start going down at that point. Since protected culture cucumbers are seedless, overmaturity is not as much of a problem as in field cucumbers, where cukes will begin to have hard seed when picked overmature. But the best eating quality in cucumbers comes after the cuke has filled out and reached its optimal length, and before it has started to bloat.

If you are growing a new variety and are unsure where the sweet spot of maturity is, undermature cukes tend to look slightly angular because they haven't filled out completely yet, like an underinflated balloon. And an overmature cuke will have stopped growing in length and started expanding outward only. Most cukes are at the right diameter when they no longer have pointy ends and are cylindrical, but before their shape has begun to inflate. Ultimately, there is some discretion in when to pick cucumbers, so show workers examples of what you want them to pick and not pick.

Post-Harvest/Storage

It's important to get cucumbers out of the sun as soon as possible, since their dark skin absorbs a lot of heat the longer they are left out. Store cucumbers at 55 to 57°F (13–14°C), and at 95 percent humidity.

As long as they are held under ideal post-harvest conditions for only a few days until sale, you can retail long European cucumbers or put them in a CSA box without shrink-wrapping. For a produce sales chain much longer than that, or wholesaling where handling is out of your control, shrink-wrapping should be used.

On a very small scale, stretchy plastic wrap may be used to wrap the cukes, with the ends twisted to seal in moisture. On a medium scale, machines are available that will speed up the process. The simplest of these can be loaded with a large roll of plastic wrap, with a hot wire to cut it. A person stands in front of the machine and wraps each cucumber, and when they lift the wrapped cucumber the hot wire cuts the plastic. On a larger scale, cucumbers are put into a loose tube of plastic wrap and then sent on a conveyor belt through a heater, which shrinks the wrap to fit the cucumber. These have a very high output and are much more expensive than the other options.

Most American slicer and pickling types do not need shrink-wrapping, because their skins are thick enough to prevent dehydration. Most of the mini/midi/cocktail thin-skinned cucumbers can benefit from some type of dehydration protection, just like long Europeans. This packaging may also be useful to keep the cucumbers together in a marketable unit, since the smaller types are usually sold in groups rather than by the piece or the pound. Shrink-wrapping cocktail or snack cucumbers to a small cardboard tray is a common industry method that also has nice presentation. Bagging small-fruited cukes for sale may be a more practical option for smaller growers. The drawback is that sometimes the bag can fog up, making the cucumbers hard to see.

Selling

Look for any way to de-commodify your greenhouse produce by promoting its thin skin and superior flavor. People may not realize that you don't have to peel thin-skinned greenhouse cucumbers. Even the American slicer types bred for greenhouse production tend to have thinner skins than those for field production.

Smaller-fruited types like picklers, minis, and cocktail cucumbers usually have to be marketed in containers. If you sell by the container, you can get a higher price per pound than consumers would be willing to pay by the piece. Any packaged produce should have a sticker that touts its benefits and helps to de-commodify it. Farm name, logo, and any attributes such as organic, pesticide-free, local, delicious, and/or produced with green energy — the sticker is an opportunity to educate the consumer about the benefits of your produce.

CHAPTER ELEVEN

Peppers

Peppers are a slower-growing and lower-yielding crop than the other greenhouse vining/fruiting crops. They can, however, be a profitable enterprise by themselves or as part of a mixed-crop greenhouse. The comparatively lower yields are partially compensated for by the fact that since they grow slowly, they also require less labor to maintain. Also, ripe peppers command a premium in the marketplace, which should help make up for the yield volume. Ripe bell peppers are the type most commonly grown in protected culture.

Fruit setting in peppers is described as flushy. It may be useful to think of pepper fruit set in terms of loading and unloading. The plant will load with fruits up to the level that the plant can support (generally three to six fruits per stem). Once the plant has set all the fruits it can support, it will abort a few flowers until some of the existing fruits are harvested, or unloaded. The amount of fruit a pepper plant can support is a function of its health. A robust, healthy pepper plant will be able to load more peppers than a weak, stressed pepper plant.

Varieties

Greenhouse peppers are an example of a crop where carefully matching the variety to the environment you plan to grow them in is very important. Peppers aren't as strong as tomatoes or eggplant; they are much slower growing and have less vigor to overcome stressful conditions.

Stressful conditions sap a pepper plant's energy. Most tomato and eggplant varieties are vigorous enough to overcome stresses – cold temperatures at night, overly hot temperatures during the day, salinity, et cetera – and remain productive. Peppers, on the other hand, have less vigor and are in danger of stalling out in adverse conditions. When grown under stressful conditions, pepper plants can slow down almost to a standstill, which results in very low yield and susceptibility to diseases and pests. I've seen this happen in unheated high tunnels in the Northeast. Because of this, I recommend that if you want to grow peppers in an unheated high tunnel in a cold, short-season area, you're better off growing a good field variety in a more standard field-growing style, like a low trellis, to keep the plants off the ground with little or no pruning.

The greenhouse growing methods and pruning style described later in this chapter under "Trellising" and "Pruning" are best reserved for heated greenhouses, or at the very least high tunnels in mild, warm-season areas. Pepper plants are divas; you must cater to them by providing the environment they like if you hope to achieve healthy, happy plants and high yields.

Greenhouse peppers are divided broadly between "high-tech" and "low-tech" varieties. High-tech varieties are also sometimes called Dutch varieties, and low-tech are called Mexican or Spanish varieties, reflecting the differences in the environments that they were bred to thrive under.

As with other greenhouse crops, the Dutch are pioneers in breeding and developing peppers that thrive under greenhouse conditions. And the way the Dutch (and others using similar high-tech greenhouses in other locations around the world) have maximized yield is by minimizing stress for the plants. Growing them at a

Ideal Pepper Climate

OPTIMAL TWENTY-FOUR-HOUR AVERAGE TEMPERATURE

A good twenty-four-hour average temperature for peppers is 70 to 72°F (21–22°C).

MIN/MAX TEMPERATURE

Do not go below 61°F (16°C) or the plants will shut down. Try not to exceed 82°F (28°C).

CLIMATE

Pepper seedlings start out as single-stemmed plants; they transition to multistemmed plants as they develop. The place where the plant branches for the first time is known as the split. When seedlings reach this stage, it's a good idea to transition from a more vegetative temperature regime to a more generative climate to encourage the plant to put energy into making peppers. Increasing carbon dioxide levels to 700 to 800 ppm and supplementing light levels if you're propagating at a dark time of year will make plants grow faster and shorten the amount of time to transplant.

Pepper plants are ready to transplant to the greenhouse when the first flower is fully developed and ready to open. For the week after transplanting, run a daytime temperature of 73 to 74°F (23°C) and a night temperature of 69 to 70°F (21°C). The high temperature and narrow day/nighttime temperature differential will promote vegetative growth and fast rooting that will help with establishment in the production greenhouse.

After the initial week in the greenhouse, bring nighttime temperatures down to 64 to 68°F (18–20°C), and maintain daytime temperatures between 70 to 75°F (21–24°C). This climate with a larger spread between day and night temperatures will be more generative and will help the plant start to put its resources into strong flower and fruit production.

Once the plants have set a full load of fruit and harvest has begun, maintain a day temperature of 70

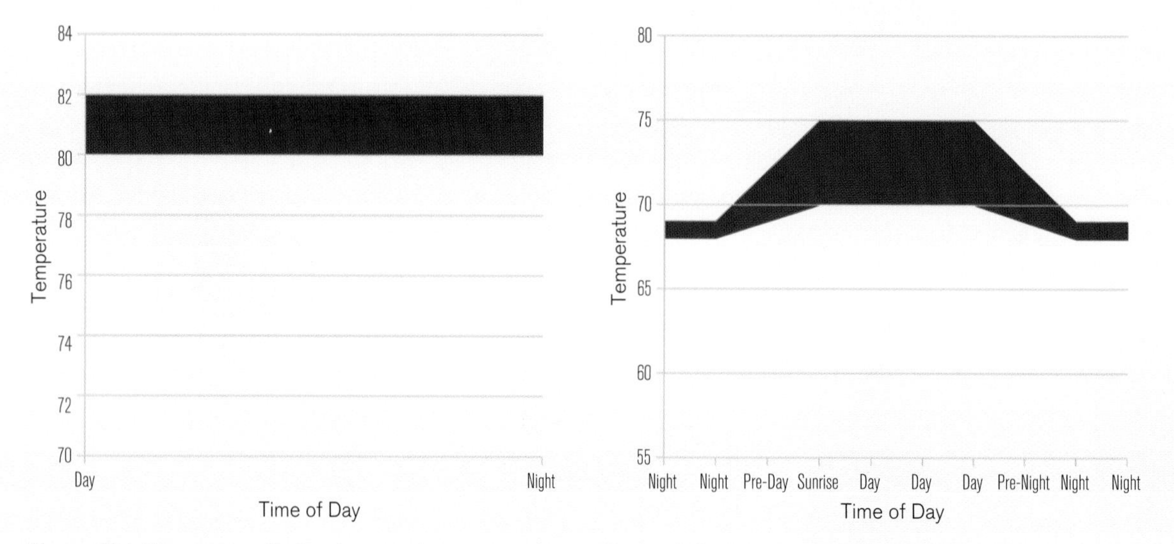

Figure 11.1. Pepper germination temperature.

Figure 11.2. Pepper seedling growth temperature.

to 74°F (21–23°C) and a night temperature of 63 to 65°F (17–18°C). Cooler night temperatures can be used to balance out hotter daytime temperatures, though even on cold days make sure to maintain a warmer temperature during the day than at night. Allowing a flat temperature profile will have a vegetative effect on the plants, and this should be avoided unless the plants need a strong push in this direction.

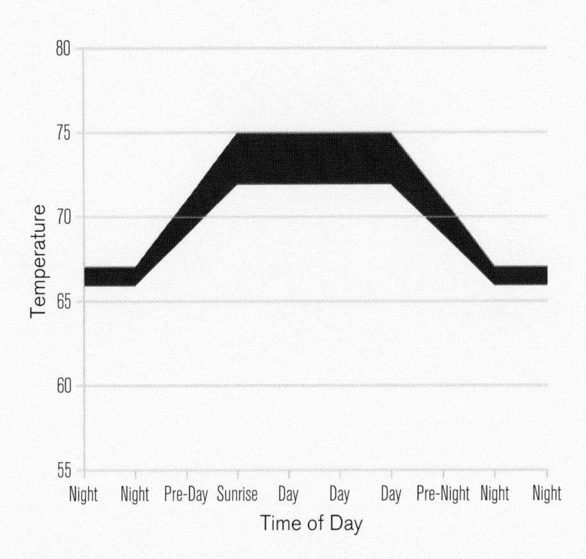

Figure 11.3. Pepper temperature for seedling growth after the appearance of the king flower.

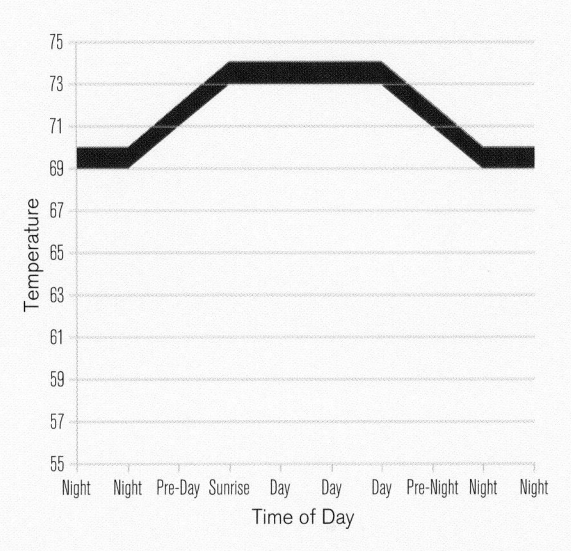

Figure 11.4. Pepper temperature for the week after transplant.

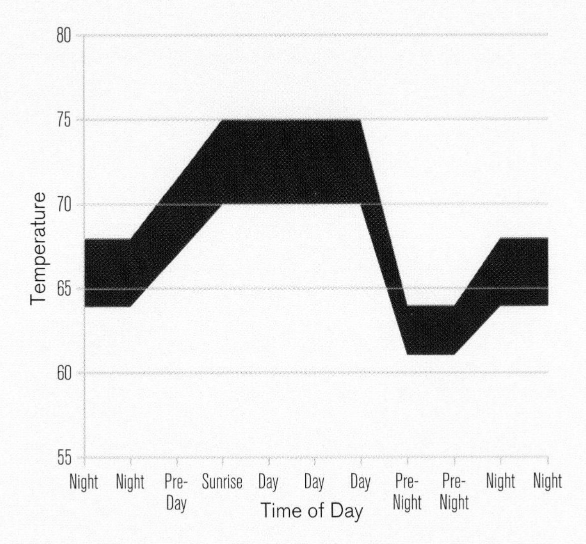

Figure 11.5. Pepper temperature at the beginning of harvest.

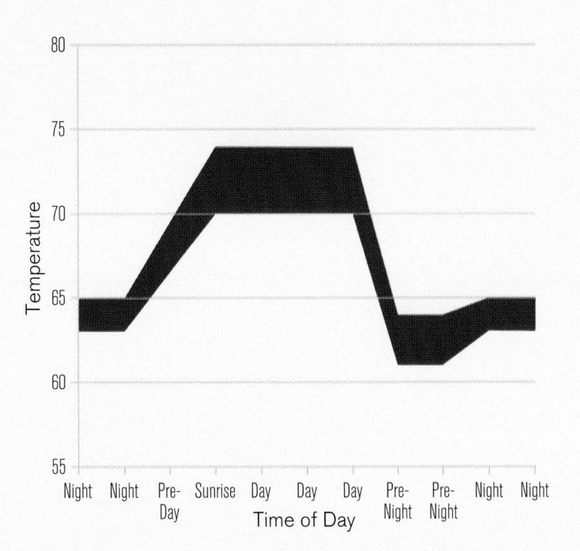

Figure 11.6. Pepper temperature at fully loaded harvest.

Figure 11.7. The equipment found in a modern high-tech pepper greenhouse. In the foreground is a harvest cart the peppers are picked into. The top can be closed to keep the sun off while the peppers are being moved to the packing area. The white tubes guide the peppers down into the harvest carts, since they may be picked from 12 feet (4 m) off the ground or more. Parked at the corner of the intersection is a scissors lift used to reach the top of the plants. Note the overhead mirror in the intersection to prevent accidents.

high temperature for the species and with as little stress as possible results in the highest yield. As discussed in chapter 7, growing fast under nonstressful conditions tends to promote vegetative growth in plants. So pepper varieties for high-tech greenhouse growing tend to be more generative than varieties for low-tech growing. Otherwise, if a vegetative variety is grown under the vegetative conditions in a high-tech greenhouse, it may become unbalanced in a vegetative manner and put too much energy on growing leaves and stems instead of fruit, and yield may suffer.

High-tech varieties that are more generative naturally put more energy into fruit and flowers at the expense of vegetative growth and plant vigor. They may be known as Mexican or Spanish varieties, because there's a lot of production of greenhouse vegetables under relatively simple plastic greenhouses in Mexico and Spain. They take advantage of climates that naturally offer good growing conditions for greenhouse crops over much of the year. Their degree of control isn't as great as in a heated, high-tech greenhouse, but with less investment in greenhouse structures and heating, their overhead is also lower while they still enjoy very high production.

These greenhouses, and simple greenhouses in general, do not have as much climate control to manage the environment as carefully as in a high-tech greenhouse. The result is plant stress when the conditions cannot be kept within the crop ideal. Since plant stress tends to sap the vigor of the crop, a more vegetative variety is used for growing in low-tech greenhouses, in order to keep it growing through the anticipated stressful conditions. It may be worthwhile trying both types in your greenhouses, but this should help explain why each type is bred for its respective environment.

In general, greenhouse peppers are bred to be less bushy, with a more regular growth habit, by branching evenly into two at each node and having a longer internode length. The vast majority of greenhouse breeding for peppers is for bell peppers, which means that if you are looking for other types to grow in a greenhouse, you may have trouble finding any with specific adaptations for protected cropping. But as greenhouse growing becomes more popular, there are more types coming on the market. In particular, long tapered Italian-frying-type peppers, medium-long cornitos, and mini snack peppers are becoming more common. Hot peppers are also starting to be bred for protected cultivation.

One thing to keep in mind when growing any of these varieties besides bell peppers is that the smaller-fruited types are much lower yielding than bell peppers. Even when density is increased for the smaller types, maximum yields may be only half of what bell peppers will yield in the same space.

This is why most of the smaller peppers are usually sold in containers or bagged by the piece, not the pound. If the price per pound were stamped on the container of most smaller peppers, they would look really expensive next to the bells. But of course smaller-fruited varieties take just as many resources and, if anything, more labor to grow, since picking the same weight takes longer with smaller fruits. So if you're going use your precious greenhouse real estate to grow them instead of bell peppers, you've got to justify it by getting a higher price per pound to make sure the space pays for itself.

Peppers from Seed to Sale

Once you've chosen your varieties, it's time to think about how to make the most of them. Below are pepper-specific recommendations to take you from seed to sale.

Propagation

See fruiting crop propagation on page 77 in chapter 6. Keep in mind that peppers are slower than the other vining crops in every way, and will take longer to germinate and get to the size for transplanting into a larger block or pot. Germinate at a constant day and night temperature of 82°F (28°C). After germination, grow seedlings at a day temperature of 70 to 75°F (21–24°C) and a night temperature of 68 to 69°F (20–21°C). Transplant seedlings from plugs or cells into blocks or larger containers when the true leaves become visible, two or three weeks after germination.

When you see the king flower, you can start a more generative temperature regime, like day 72 to 75°F (22–24°C), night 66 to 67°F (19°C). This will encourage the plant to put energy into the flowers once it's producing them.

Once the king flower is open, the seedlings are ready to be transplanted into the production greenhouse. Just before or at transplanting is a good time to pinch the king flower off, as it should not be allowed to develop into a fruit. For the week after transplanting, maintain a warm, narrow temperature profile of 73 to 74°F (23°C) day, 69 to 70°F (21°C) night to encourage rapid rooting and vegetative growth.

In the early stages of harvest, use a day temperature of 70 to 75°F (21–24°C) and night temperature of 64 to 68°F (18–20°C). During full harvest, once you have four to six peppers developing per stem, you can use a more generative temperature scheme of 73 to 75°F (23–24°C) during the day and 63 to 65°F (17–18°C) at night.

Peppers are sensitive to rapid temperature changes, so keep in mind not to change temperatures faster than 2°F (1°C) per hour when possible. A two- to three-hour pre-night treatment at or close to the nighttime minimum is helpful for good flower and fruit production. If you need cooler temperatures to balance hot daytime temperatures, or to balance excessively vegetative plants, you can set pre-night temperatures as low as 61°F (16°C). Some growers will use a faster-than-usual temperature change of 2°F (1°C) every fifteen minutes if they need a really strong generative action. Don't go below 61°F (16°C), as it will cause the plant to shut down until it warms back up.

Pepper plants will have difficulty setting fruit if night temperatures stay above 68°F (20°C). In areas with average night temperatures warmer than this, pepper flowers may drop off without forming fruit unless the greenhouse can be cooled or until night temperatures come down. Getting temperatures this low at night in the summer can be a challenge in many areas.

Crop Cycles and Timing: Northern North America

Since peppers are such a slow-growing crop, heated-greenhouse growing uses a very long season in order to maximize the amount of productive time in relation to time spent in propagation and the seedling stage. In the upper United States and southern Canada, where most greenhouse peppers are currently grown in North America, the most common cropping schedule is to plant seeds in October, transplant into the production greenhouse in the first half of December, and harvest peppers from sometime in March until November or early December.

This is where the long season to make up for the lengthy propagation period comes in. Note that seeds planted in October don't yield marketable peppers until five months later in March! That is a long time to have costs associated with pepper production without any yield. Granted, part of that time is spent in a propagation greenhouse at high plant densities so costs are lower. And part of the reason the plants are growing so slowly is because they are growing through the depths of the winter. But that emphasizes why you need an eight-month harvest season: to make up for five months of plant growing before any yield.

At this point, no one is growing peppers year-round with fruit through the winter months in northern North America. There is not enough light to carry a pepper fruit load over the winter months. And the value of peppers that would be harvested through the winter does not justify the expense of lights.

Crop Cycles Closer to the Equator

For locations where extreme heat in the summer poses more of a production problem than cold temperatures in the winter, the above crop cycle may be flipped. Instead of being out of production during December, being out of production during July or whatever the hottest month is may be a viable option. Plants may be propagated in shade houses during the hottest part of the year, and can go into the production greenhouse as soon as the hottest weather is past.

One important factor to work around is that peppers do not set fruit well when nighttime temperatures are above 68°F (20°C). If night temperatures are above this level for extended periods of the year, this will likely interfere with fruit set and make pepper production difficult, if not money-losing. At the very least, ask a pepper grower in your area if this is a problem or do a test planting with another crop you know can be successfully grown. In many areas closer to the equator, pepper crops may be grown through the winter because day length varies less the closer you get to the equator, until day length is nearly the same year-round at the equator.

Grafting

Though rootstocks are available for peppers, breeding has not gotten to the point where much grafting is done in commercial pepper production. Rootstocks provide protection against some soil diseases but just like cucumber grafting, yield boost has yet to catch up. If the breeding ever gets to that point, I'm sure pepper grafting will become as commonplace as tomato grafting is now.

Transplanting

Transplant to the production greenhouse when the king flower is fully developed, but do not let it turn into a fruit or the plant may be permanently stunted. The plant needs to get larger before it can support a fruit load.

Spacing

Standard spacing for two-headed greenhouse peppers is 3.5 plants per square meter, for a final density of 7 heads per square meter. Growers using four heads per plant will transplant half the number of pepper plants and develop twice as many shoots to get the same density of seven heads per square meter. This results in a very dense hedge-like pepper row – but keep in mind that peppers are slow growing, have small leaves, and won't be lowered and leaned, so the individual vines don't need to be separated.

This density results in the highest yields, and is achieved by anchoring a vine every 6 to 7 inches (15–18 cm) or so along our standard 2-foot-wide (60 cm) double row/3-foot-wide (90 cm) walkway. For two-headed plants this is achieved either by planting a single row with plants 6 to 7 inches apart in a V-system, or by planting double rows of peppers 12 to 14 inches (30–36 cm) apart, with both heads anchored to the same wire about 6 to 7 inches apart. If you'd like a lower planting density, you can space vines out more along the wire.

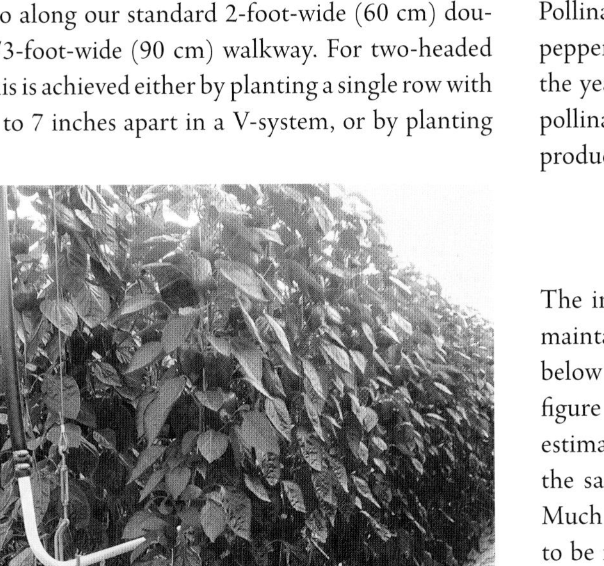

Figure 11.8. A double row of bell peppers in a heated greenhouse. The white pipe visible between the rows provides heat within the canopy, which helps keep the leaves dry and fruits warm for faster ripening.

Smaller-fruited varieties of pepper should be grown with three stems per plant at the same planting density as a bell pepper with two stems, so planting at 3.5 plants/m^2 results in 10.5 stems/m^2. Even with this high planting density, most peppers smaller than bells can result in as little as half the yield of bell peppers, with the smallest cultivars yielding the lowest.

Pollination

Pollination by bumblebees is usually unnecessary for peppers. If fruit set is really poor at difficult times of the year, like winter, some growers use bees to improve pollination. But bees are not usually used in pepper production.

Labor

The industry estimate is that an experienced crew can maintain an acre of greenhouse peppers as described below in about a hundred hours per acre per week. This figure is probably most useful in comparison with the estimate that a similarly experienced crew can maintain the same amount of tomatoes in 140 hours per week. Much of this is due to the fact that peppers only need to be maintained every other week, whereas the faster-growing tomatoes, cucumbers, and eggplant will need at least weekly maintenance. I say it is most useful in relative terms because in reality, the time it takes different farms to maintain the same amount of crops is all over the place. Still, knowing what the industry standard is for the most efficient greenhouses can be useful in knowing what to shoot for.

Trellising

Because their stems are brittle, peppers are not usually lowered and leaned. They are usually grown up a piece of twine that is tied to an overhead wire, which may need to be as tall as 17 feet (5 m) for an eleven-month heated

Figure 11.9. A crop of greenhouse bell peppers. The plants are approximately 10 feet (3.3 m) tall. You can see the pipe rail (*right*), which is used for heating and to guide scissors lifts and harvest carts.

pepper crop. Peppers can be grown in a V-system, with the plants between two overhead wires and half the leaders going to one side, half to the other. Peppers can also be grown so that both heads of each plant go to the same wire. In a single-row layout, this means every other plant goes to every other side. In a double-row layout, it means that all the heads of the plants on the left go to the left wire, and the opposite for the plants on the right.

Pruning

Bell peppers trellised in protected culture are typically kept to two or four leaders. Smaller-fruited cultivars, like hot peppers, snack peppers, or long sweet peppers, are planted at the same density as bells and allowed to develop three heads to compensate for the lower yields of smaller-fruited cultivars by increasing plant density. This is possible with smaller peppers because these types tend to have smaller leaves, and can be crammed in tighter.

Even in a heated greenhouse, due to their slower growth, pepper plants only need to be pruned every two weeks, or after 6 inches (15 cm) of growth or developing two or three new shoots, if they are growing more slowly.

The more heads a pepper plant has, the smaller the fruit will be. Four heads are common in Holland, because larger fruit isn't as important, but two heads per plant is most commonly used in North America because there is a premium for larger fruit.

The first flower on a pepper plant is formed at the split where the plant branches into two for the first time. This king flower is always removed to allow the pepper to put its energy into growing the plant.

Each node on a pepper plant is called a level. The split, or crotch, of the plant where the king flower is formed is referred to as level zero. The node above level zero is called level one, and so on and so forth on up the plant. Up to level zero, the pepper plant has grown as a single stemmed plant. The appearance of the king flower signals a change from a vegetative, seedling growth habit into a more generative, mature growth habit. At the first flower the plant will go from single-stemmed to multistemmed.

At every node after the split, there will be a leaf and a flower, and the branch will split into two. If you want to maintain two heads on the plant, the first split is the only one where you will keep both branches. At every other split after that, you will be keeping one branch and eliminating the other. If you want to maintain four heads on the plant, you will keep both branches at level zero and at level one, for a total of four, and then keep one branch and eliminate one branch at every level after that.

The first rule of pepper pruning is not to work on the top 8 inches (20 cm) of the plant. You may be eager to prune and get right up into the head of the plant. But the top 8 inches isn't well developed, and it can be difficult to tell what to cut and what to keep that close to the top. Also consider that when plant parts are still that small, they're vulnerable to damage by insects. You want to

Figure 11.10. The view from the top of a large pepper greenhouse. The yellow cards are used to monitor pests in the greenhouse. The number and species of insects stuck to the cards is counted on a regular basis.

let them grow out until you can get a better look at the branches before making a decision about which one to keep. If one branch is weak or abnormal, this will allow you to eliminate it and keep the good branch.

Top the plants six or seven weeks before you remove them from the greenhouse, to stop them from devoting energy to vegetative growth and help them focus only on maturing the fruits that are already set.

Flag Leaf

Pepper fruits are sensitive to sunburn. One pruning strategy to develop a better canopy is called leaving a flag leaf. When you are making the decision at each level of which branch to keep and which to eliminate, instead of pruning off the whole branch, you can top the branch just above the first leaf it produces (see figures 11.11 and 11.12). Place your fingers or clippers between the leaf on the bottom and the flower and two branches above it. In one cut you can eliminate both branches and the flower, leaving just the leaf at the next level. This is called the flag leaf, the only thing left at the tip of a branch that was terminated.

Fruit Pruning

The first flower on the pepper plant, the king flower, is almost always removed to allow the plant to develop more leaves before it has to support a fruit load. Once the king fruit is removed, you must decide how soon to begin allowing fruit to set. This depends on time of year and crop conditions.

For long-season pepper crops that are transplanted in December or January, most growers do not allow flowers to turn into fruit until the second or third level. There is so little light available at that time of year, early fruit set may come at the expense of the plant in the long run. Flowers need to be removed by pinching or clipping to keep them from developing into fruit.

The earliest fruit is often the most profitable, so growers are always tempted to start leaving fruit earlier to get an early fruit set. Some growers with optimal conditions may allow fruit to set as early as the first level if they think their plants can handle it. This is more common in later-spring-planted pepper crops, as the plants are more able to handle an earlier fruit set when light levels are higher and growing conditions are better.

One common strategy is to leave one pepper at each level of each branch of a pepper plant. This doesn't usually involve removing any fruit, because bell peppers

Figure 11.11. Here's how to make the pruning cut to leave a flag leaf on greenhouse peppers: On a branch that is to be terminated with one leaf, put the clippers between the leaf on the bottom, and two branches and developing flower bud on top.

Figure 11.12. Make the cut, severing the flower bud and two developing branches, leaving only a leaf on the end of the branch.

almost always grow one flower per level. This does tend to lead to flushy production, as the plant may set four to six peppers in a row, become overwhelmed with fruit load, and then abort a few.

An alternative pruning strategy is to leave two flowers on one stem and one on the other, and alternate between two flowers in a row and just one for each branch. This is sometimes called two-in, one-out pruning, because you leave two flowers and take one out on each vine all the way up the plant. It's a little bit more labor, but it will produce the same overall yield as leaving one flower per level per branch. And it should eliminate most flushiness and make yields more consistent week-to-week.

Pests and Diseases

See appendix B for a general overview.

European Corn Borer

European corn borer (ECB) larvae usually bore into pepper fruit through the calyx. This is sometimes visible by a small hole and frass on top. When it goes undetected, a pepper can look fine from the outside but have a worm on the inside. This is one pest that tends to be reduced in peppers grown in protected culture. The moth that lays the ECB egg is not as likely to make its way into a covered structure. Peppers that have ECB damage cannot knowingly be sold because there is likely interior rot with a worm inside.

Harvest

Harvest ripe peppers when they are 70 percent or more colored; they will ripen the rest of the way off the plant. Peppers that are harvested with less color may ultimately reach full ripeness, but ripening may be slow, which will reduce shelf life. To determine ripeness for harvest of green peppers, look for full-sized fruit that have thicker, firmer walls than unripe fruit. Peppers gain size before wall thickness.

In some varieties the shade of green may change slightly when it goes from immature to mature. Otherwise you can tell mature full-sized fruit from undermature full-sized fruit with thinner walls by giving them a gentle squeeze. It would take too long and possibly damage the fruit to squeeze every single pepper. But if pickers aren't sure what to harvest, you can use this as a test to help them get the idea.

Use clippers or a knife to cut peppers off flush with the vine. This leaves a long stem on the pepper. Just like pruning stubs can be a point of entry for pathogens, stubs from harvesting can be vulnerable as well. If the pepper wasn't cut off cleanly in the first place, it's worth cleaning up the stub on the plant so it doesn't get diseased.

Flavor

The main factor in pepper flavor is ripeness. Peppers that have ripened will be sweeter and have a more complex flavor than green peppers. Though there is variation among varieties, pepper flavor isn't as variable as some other crops. If you are specifically growing green peppers, some varieties are bred to ripen very slowly and stay green for a long time. So there is less of a chance that they will inadvertently ripen if you don't harvest them right on time. Otherwise you can just harvest an unripe regular colored variety.

If you're growing hot peppers, note that the amount of heat varies widely depending on growing conditions. Peppers under any kind of stress will tend to be hotter than peppers grown under optimal conditions.

Yields

The highest yields in North America are being obtained by growers with high-tech greenhouses, using the hybrids well suited to their growing conditions and style. Yields in the neighborhood of 32 kg/m^2 are being recorded in commercial greenhouses, which translates to something like 20 pounds (9 kg) per plant or 10 pounds (4.5 kg) per leader. Medium-fruited cultivars may yield up to 23 or 24 kg/m^2, and small-fruited varieties may only yield 16 to 18 kg/m^2. Though this is only possible with heat and good climate control, it's worth knowing since it sets the mark for what is possible.

Post-Harvest/Storage

Peppers bred for greenhouse production tend to be fairly uniform, but some wholesale markets demand a greater degree of uniformity. If so, grading and sizing may be necessary. Store peppers at 45 to 47°F (7–8°C) at 95 percent humidity. Do not store peppers with ethylene-producing fruits and vegetables; they will cause peppers to ripen faster and reduce shelf life.

Selling

Ripe bell peppers are usually sold wholesale by the box, or retailed loose. "Stoplight" packs with a red, yellow, and green (or orange) pepper packed together have become a popular way to market peppers. Smaller-fruited types usually need to be sold in a pint container or bagged by volume to get a high enough price to justify growing.

CHAPTER TWELVE

Eggplant

Eggplant is a distant fourth in greenhouse fruiting-crop production. For one thing, less eggplant is consumed than the other crops, so the market is smaller. One of the reasons so little eggplant is grown is because the big growers, who produce the largest volume of greenhouse vegetables, go after the biggest markets: tomatoes, cucumbers, and peppers.

Eggplant can punch above its weight when it comes to its value for small- and medium-sized growers, however, and be a high-yielding and productive crop. For growers who need to maintain diversity in a market stand or CSA box, eggplant can grow well with the other fruiting crops and provide something different. For larger growers, having eggplant to offer may make your assortment stand out to wholesale buyers.

The fact that eggplant is the least grown of the four main protected fruiting crops means that there is less of everything for the crop. It isn't as well studied as tomatoes, cucumbers, and peppers, so there is less information and fewer protected varieties for eggplant. Luckily, eggplant culture is very similar in many respects to tomato growing. Much of the wealth of information on tomatoes can be applied to eggplant, with a few important differences.

Varieties

There is less breeding underway for greenhouse eggplant than for any of the other fruiting crops. This is simply a function of less greenhouse eggplant being grown than the other crops, so corresponding amounts of resources are dedicated to breeding. But along with the general increase in greenhouse growing, there are more eggplant varieties coming onto the market.

The breeding goals of greenhouse eggplant include adaptability to a wide range of temperature conditions, since the temperature spread may be wider in a greenhouse than outdoors. Plants are selected to be taller and less bushy than field varieties so they can be trellised at high density and still have good airflow. Some field varieties have spines on the leaves and calyxes. Greenhouse varieties have been bred to have no or very reduced spines. It is much nicer working on spineless eggplant. Besides being less comfortable for the workers, eggplant that have spines on the calyxes stand a good chance of puncturing each other when they are harvested, reducing quality and shelf life.

Most varieties that have been developed for the greenhouse are of the black Italian type, because that is the most popular in Europe and North America. They have dark purple skin that is almost black, with a blossom end that's larger than the stem end. There are also some white cultivars with the same shape, and some purple-and-white-striped ones, which are sometimes marketed as graffiti eggplant regardless of variety; long, cylindrical varieties that are popular in Asia, in shades of purple, green, and other colors; and even some varieties that look like Rosa Bianca and other Italian heirlooms.

Ideal Eggplant Climate

OPTIMAL TWENTY-FOUR-HOUR AVERAGE TEMPERATURE

Eggplant have temperature requirements very similar to tomatoes', and when grown together the ideal temperature program for tomatoes works well for eggplant. Eggplant will produce the maximum amount of fruit when the average temperature is 66°F (19°C).

MIN/MAX TEMPERATURE

Try not to go below 55°F (13°C) or over 86°F (30°C).

CLIMATE

Run the greenhouse temperature flat day and night at 75°F (24°C) for the first week after transplanting. Running the greenhouse temperature warm and flat will encourage rapid rooting in and vegetative growth. This can be particularly helpful to get the transplants established, since eggplant roots aren't as strong tomato roots. Gradually lower the temperatures by 2°F (1°C) per night until you reach a daytime temperature of 72°F (22°C) and a nighttime low of 66°F (19°C).

As the plants reach maturity and become fully loaded with fruit, use a pre-night treatment and a temperature regime with a wider differential to have a more generative effect and keep the plant's energy

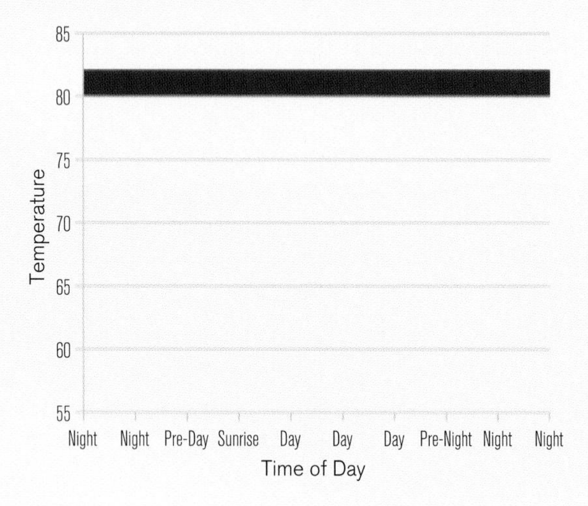

Figure 12.1. Eggplant germination temperature.

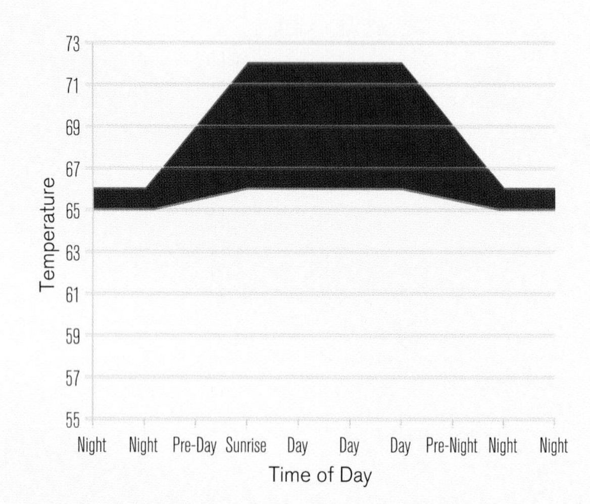

Figure 12.2. Eggplant seedling growth temperature.

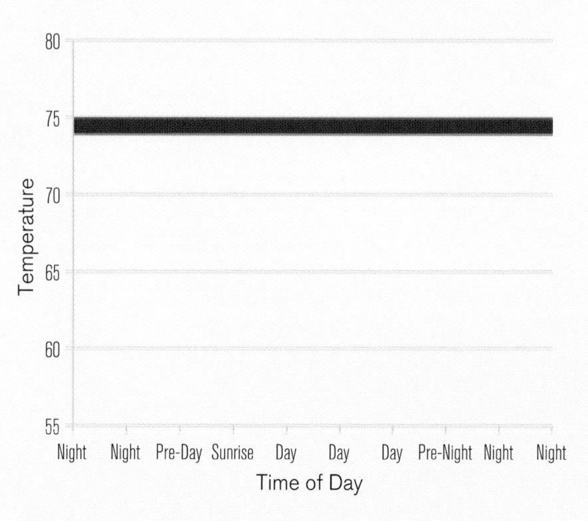

Figure 12.3. Eggplant temperature for the week after transplant.

directed toward fruit production. Maintain daytime temperatures around 68 to 72°F (20–22°C), and lower night temperatures to 63 to 66°F (17–19°C) to achieve a greater day/nighttime differential.

Eggplant are very vigorous and vegetative by nature. Pre-night treatments similar to tomatoes, down to a minimum of 60 to 64°F (15–18°C), will help keep flowering and fruit setting strong. During hot weather when the daytime high is going above the ideal, keep the plants at a lower nighttime temperature for longer to counter the effect of the hot weather on the average temperature.

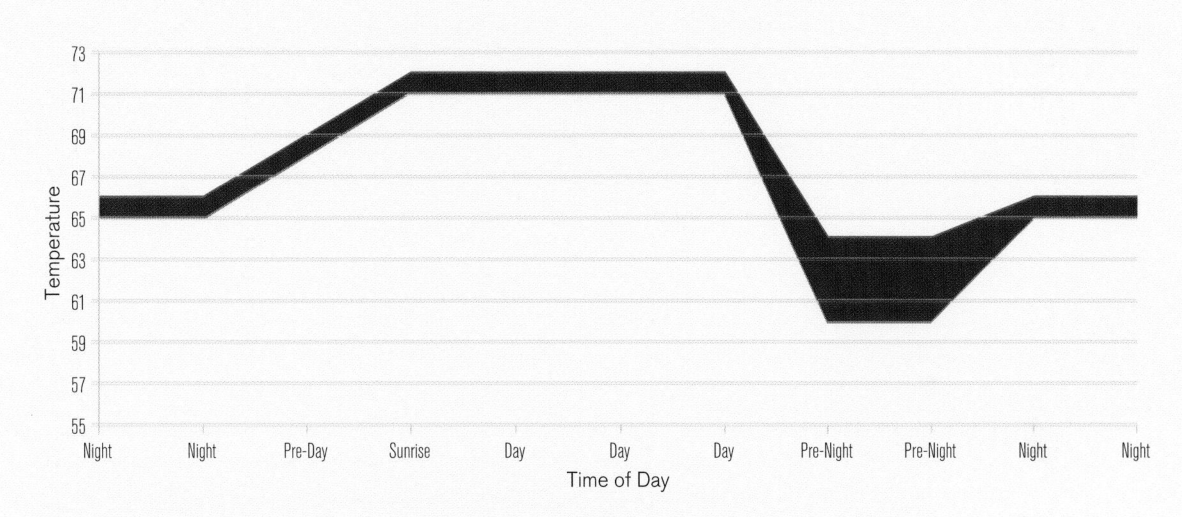

Figure 12.4. Eggplant temperature at the beginning of harvest.

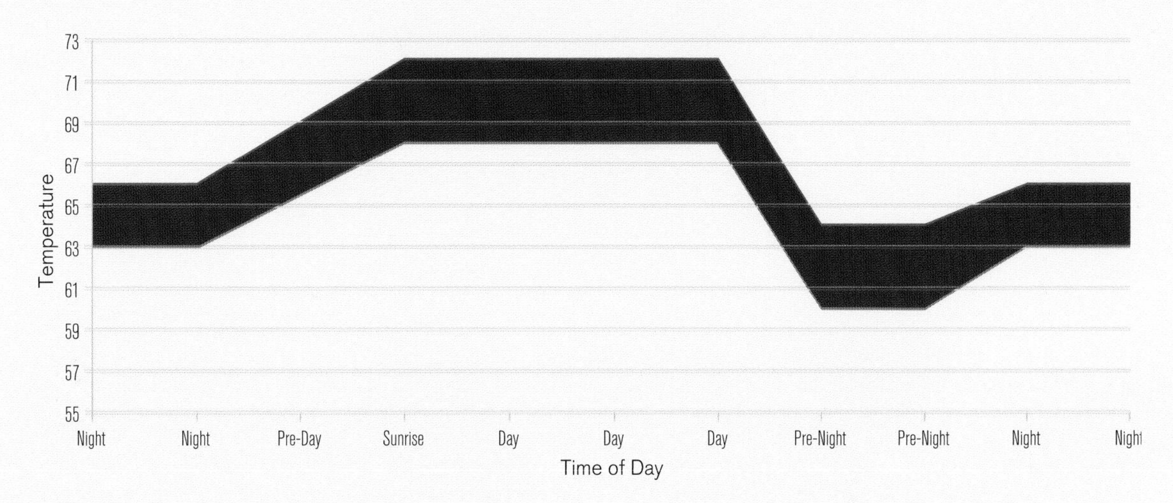

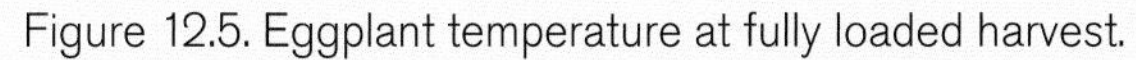
Figure 12.5. Eggplant temperature at fully loaded harvest.

Eggplant from Seed to Sale

Once you've chosen your varieties, it's time to think about how to make the most of them. Below are eggplant-specific recommendations to take you from seed to sale.

Propagation

Sow seeds into the medium of your choice and germinate at 80 to 82°F (27–28°C). Eggplant are slower germinators than tomatoes, with emergence taking place six to eight days after sowing. After emergence, use a 66 to 72°F (19–22°C) daytime temperature and a 65 to 66°F (18–19°C) night temperature. Transplant into a larger block or pot two to three weeks after emergence, before the seedlings start shading one another and causing stretching. Eggplant are commonly grown until 10 to 12 inches (25–30 cm) before transplanting. Plants much larger than this can be unwieldy to transplant, and would require even larger blocks or pots.

Crop Cycles and Timing

Eggplant can be a very long-season crop in heated greenhouses, with growers in northern North America using a yearlong growing cycle. One common cropping cycle in areas where summers are not prohibitively hot for production is to start plants in November or December, transplant to the production house in mid- to late January, begin harvesting in March, and continue harvesting until late November or early December.

In this model, growers start seeds for the next crop as they are terminating the old one. This leaves the time from termination of the old crop until the planting of the new one, or roughly from late November/December until January, to clean out the greenhouse and get it ready for the new crop. In areas with very hot summers, or in low greenhouses that aren't tall enough to grow a long-season crop, a spring and a fall crop can be planted. In this scenario, the spring crop could be planted in November, transplanted to the greenhouse in early January, and harvested until July, or whenever the weather gets too hot or the plants get too tall to continue growing. The fall crop could be planted in late June or early July, transplanted into the greenhouse in August, and harvested from late September or early October until late November or December.

Grafting

Eggplant is just as easy to graft as tomatoes, and many of the same considerations play into a decision about whether or not to graft. One thing that makes grafting attractive is the fact that there is very little in the way of soilborne disease resistance available for eggplant. Open up any seed catalog and you will see that there are very few disease-resistant varieties. So if you have soil diseases, grafting may be one of the only ways to overcome them.

Eggplant can be grafted onto any tomato, and in fact most of the commercially available eggplant rootstocks are tomatoes. The challenge is to find a good combination of top and bottom for your circumstances. Depending on what is available, there may be tomato rootstocks that are specifically recommended for eggplant. There are also some eggplant rootstocks for eggplant.

Though eggplant rootstocks don't have to be specifically used for eggplant, many of the rootstocks that match well tend to be of the lower-vigor, more generative type. One important consideration in choosing an eggplant rootstock is to keep in mind that eggplant tends to naturally be a very vigorous, vegetative crop. So finding a rootstock variety that helps balance and doesn't result in an overly vegetative plant is important. Some of the very high-vigor rootstocks may result in excessively vigorous, leafy plants.

Timing Propagation for Grafting

Since eggplant seeds are slower to germinate and grow than tomatoes, the best way to match up eggplant scions with tomato rootstocks is to plant the eggplant first and wait for them to germinate before planting your tomato rootstocks. That way, the tomatoes will germinate and catch up with the eggplant in the correct window for grafting. Eggplant rootstocks can usually be started on the same day as eggplant tops, but as usual it's worth doing a test planting to compare the growth rate of top and bottom to make sure they match. Otherwise, eggplant grafting follows the same procedure as detailed in "Solanaceous Grafting" in chapter 8.

Transplanting

Eggplant are ready to be transplanted when they are 10 to 12 inches (25–30 cm) tall and have some flowers that are open or ready to open. It may take take eggplant seven to ten weeks to reach this size; add two weeks if you use grafted plants. A transplant of this size will likely need a support stake to remain upright.

Pollination

Eggplant do not need pollinators. As with peppers, sometimes growers experiencing poor fruiting with eggplant during the winter may be able to improve fruit set by using bees. But bees are not typically used in eggplant production.

Trellising

Eggplant in protected culture can be grown in a tall basket weave (Spanish style), or each leader can be grown up a string using a trellis clip or winding the string around each stem as with other vining/fruiting crops (Dutch style). Eggplant can grow as fast and get just as tall as tomatoes in protected culture, so make sure there's enough vertical space, depending on the length of the season.

Basket weaving can be used for growing eggplant. However, you can only grow them as tall as your stakes, and eggplant can grow up very quickly. Plus, the fruits can easily get trapped and deformed in the strings of the weave, and plants are harder to prune inside the strings that form the weave. Yet this is still a viable option for some growers who may want to limit the amount of labor that goes into a greenhouse eggplant crop. Basket-weave as you would any other greenhouse crop, and prune to your desired number of stems.

Eggplant can be lowered and leaned, but it's more difficult than with tomatoes or cucumbers because the stems are more brittle. If you want to lower and lean eggplant instead of trellising the plants straight up, start the plants with a lean in the direction you want them to go. That way when you go to lower and lean, the stems are already going in the right direction and they are less likely to break. Instead of being trellised vertically up and down, slant the vine one position over from vertical in the direction it will be leaning. In other words, anchor the vine over its neighbor in the direction the plants will be going.

If you aren't going to lower and lean your eggplant, which is most likely, tie strong twine to the overhead wire at your desired spacing, and tie the other end of the twine to the base of each plant, or attach it with a trellis clip. Otherwise, trellis as described, maintaining the plants every week, or as needed to keep them from flopping off the string.

If you don't want to lower and lean but need to keep the plants a manageable size, see "Pruning" on page 194.

Spacing

Eggplant are grown at densities anywhere from 3.6 stems per square meter up to 8 stems/m^2. The lower densities are more appropriate for unheated hoophouses, and the higher densities work better in higher-tech structures with better ventilation and heat.

In our standard greenhouse layout with double rows 2 feet (60 cm) apart and 3-foot (90 cm) walkways, 3.6 vines/m^2 can be achieved with a single row of eggplant in the middle of the row a foot (30 cm) apart. Develop two heads on each plant, so you end up with a vine every foot along each wire. You can achieve the same density using double rows with plants 2 feet apart in the row by developing two heads on each plant, resulting in one plant every foot along each wire.

The easiest way to reach the desired density of five to eight stems per meter with four-stemmed plants is by using the same planting densities as above and developing additional stems on each plant as needed to increase density. For example, going from two stems to three

Figure 12.6. A young eggplant crop being grown hydroponically in cocoa coir slabs. Note the dripper stakes pushed into the 4-inch (10 cm) blocks, which are fed by the black pipe in the center of the double row.

stems on each plant would bring the density up from 3.6 stems/m^2 to 5.4 stems/m^2. Developing a fourth stem on every plant would bring the density up to 7.2/m^2. In-row spacing can be adjusted upward or downward to give a higher or lower density. Greenhouses with better ventilation and light penetration may use a higher planting density.

Labor

Based on the industry standard of using an experienced, efficient crew, a Dutch-style trellised eggplant crop requires approximately 130 hours of labor per acre per week to maintain. Your mileage may vary.

Pruning

Due to eggplant's vegetative nature, splitting the energy of each plant into at least two leaders is recommended. Eggplant grows with a single main stem, with suckers developing at the node where each leaf is attached. Let the plant grow as a single stem until you see the first flower cluster. The sucker below the first flower cluster is usually the strongest sucker on the plant and the best one to match the main stem. If you want to grow two-headed eggplant, the sucker below the first flower cluster is the only one you should allow to develop on the whole plant. Let that one develop into your second leader. If that sucker is damaged or accidentally snapped off, the next

Figure 12.7. The top of a greenhouse eggplant. The main stem is distinguishable to the left by the twine wrapped around it. A sucker is growing out of the node at the bottom of the frame. This sequence will show you how to top a sucker in order to leave the fruit.

one up the plant (or any other sucker) can be allowed to develop into a second leader.

If you have a low greenhouse, eggplant grown with four stems will reach the top wire more slowly than plants grown with two heads. If you want more than two heads on your eggplant, pick any other strong-looking (large, vigorous) suckers close to the head of the plant to develop into additional heads. You want suckers near the top, since they're already close to the head you're trying to match. Suckers grow faster when they receive more light. Suckers from farther down the plant may be in too much shade to develop properly. Since they aren't receiving much light, they will grow slowly and never catch up with the main stem of the plant.

Once you have your main leaders on each eggplant, it's important to trellis in a timely manner in order to keep each stem attached to the string without flopping. As the leaders grow up, you will want to trim the side shoots that develop at each node, but do not remove them immediately.

Unlike the case of tomatoes, you don't want to remove the entire sucker; each shoot needs to develop until flowers are visible. Eggplant shoots produce flowers more

Figure 12.8. Place the clippers just above the first flowers to develop on the sucker.

Figure 12.9. When you make the cut, the only thing left on the sucker should be the flowers.

quickly than tomato shoots. Whereas a tomato shoot usually puts out three leaves before making a flower cluster, eggplant flowers are frequently the first thing produced as a sucker grows out. Let the sucker develop until you see a flower cluster, and then use pruners to cut the sucker off just above the flower cluster. As with all pruning cuts, cut just above the part of the plant you're keeping (flowers) in order not to leave a stub that could get infected. In this case, cut the sucker off flush with where it attaches to the flower stem.

Eggplant is prone to setting multiple flowers in a cluster. Some of these flowers will be much bigger than others. Some growers remove all flowers except the largest one, which will grow and develop faster than smaller flowers. Instead of cluster thinning as with tomatoes, I recommend leaving all the blossoms on eggplant. That way there are extra flowers in case one of them gets damaged.

Figure 12.10. In this picture, at least half the fruits are set on suckers that have been topped. You can see how much this helps increase the fruit load.

Even though smaller blossoms develop more slowly, they frequently get to marketable size, albeit more slowly than larger blossoms. They may start developing more quickly once the largest fruit in the cluster is picked, freeing up energy for the smaller blossoms. Continue pruning the plants in this manner every week or so until the end of the crop.

You will usually want to use this pruning method to set as much fruit as possible on an eggplant. This is in contrast with the fruit pruning you may do on cucumbers or tomatoes. Leaving all the fruits that form on the main stem as well as the suckers will help keep the plant balanced, and increase yield greatly. In figure 12.10, for example, more than half the fruit have grown from flowers on side shoots. If all those suckers were removed entirely, yield would be halved, and the plant might well devote the energy that is now going to fruit to excessive vegetative growth instead.

As the first fruits approach harvest stage, remove the leaves from below the lowest fruits as you would tomatoes. This will increase airflow at the bottom of the crop and make harvestable fruit easier to find. As you pick upward on the vines, continue clearing out the leaves from below the fruit to maintain airflow.

Pruning for a Short Greenhouse

If you have a really low greenhouse, you can use an alternative pruning strategy to keep four-stemmed eggplant really low. Start by growing each plant with four stems. This will naturally keep the plants lower by dividing the energy from the roots among all four heads.

Starting about a month and a half after transplanting, top the plants, snapping or cutting the growing points off each head a few inches (cm) below the top of the plant. This is exactly what you don't want to happen under normal circumstances, as it interrupts the plant's growth. It's what keeps the plant low by interrupting its upward progress. It does set the plant back a bit, but it can help you make use of a greenhouse that's otherwise too low for an eggplant crop. Eggplant's strong, vegetative nature helps it power through and stay productive.

To use this method, do not sucker the plants during the week before you intend to snap the heads. If the

Figure 12.11. This is an older greenhouse with a low top-wire height of around 6 feet (2 m), and the eggplant crop will be topped periodically to keep it low.

plants have been freshly suckered, there will be no growing points to act as an outlet for plant growth, which may trigger suckering from the base. This is not desirable, because those suckers will be down in the shade. Even though you want to interrupt the plant's growth, you still want it to come from the top. When you leave the suckers you otherwise would have pruned the previous week, those will become the new heads.

A week or two after you have removed the heads, when you normally would do the routine pruning and suckering, make sure you leave the strongest sucker per stem to develop into a new head for that stem. All the other suckers beside the strongest one can be topped to leave a flower cluster as usual.

Pests and Diseases

Eggplant is a favorite of many different pests. All the insects that like the other vining crops – thrips, whiteflies, and aphids – really like eggplant, too. In addition, Colorado potato beetles and flea beetles may attack eggplant.

Most of these insect pests prefer eggplant over tomatoes and the other fruiting crops. For this reason, and since eggplant coexist so well with tomatoes, some tomato growers may throw a row of eggplant in with tomatoes more for pest management reasons than anything else. They know that eggplant will function as a sentry crop; since pests prefer it, they will show up there first. Growers can thus monitor the eggplant to see what pests are in the greenhouse. It can also function as a trap crop, drawing pests away from the tomatoes. Beneficials can be applied at a higher rate on the eggplant to combat the higher concentration of pests that should be gravitating there.

Blossom End Rot

Eggplant may suffer from blossom end rot, though it's not as common as in tomatoes and peppers. Make sure adequate calcium is available to the plants, and follow the steps detailed in appendix B to avoid BER.

Harvest

Since eggplant is not harvested at biological maturity—that is, ripeness—like tomatoes and colored peppers, when to harvest is up to you. Two common sizes for greenhouse eggplant to be harvested at are 0.5 pound (225 g) for "baby" eggplant and 0.25 pound (113 g) for "mini" eggplant. It could be harvested the size of a Ping-Pong ball if you had a market for that, though yield would be proportionally lower the smaller it is harvested. You would just have to get a premium for "Ping-Pong" eggplant, as with grape tomatoes.

One advantage of harvesting at any of these sizes is that eggplant does not develop hard seeds until it gets larger, which is one of the biggest detractors from eggplant flavor. Any eggplant harvested before biological maturity will have better flavor than mature eggplant, which can become bitter with hard seeds and thick skin.

The eggplant fruit stem is strong and firmly attached to the plant, so it's necessary sever the fruit from the plant. Bypass pruners or harvest scissors work well for this job. Cut the stem of the fruit off flush with where it meets the vine, so as not to leave a stub that could be an entrance for pathogens. Remove harvested eggplant from the greenhouse as quickly as possible. Dark-colored fruits absorb heat very quickly in a sunny greenhouse, reducing post-harvest quality.

Flavor

Eggplant does not have the same diversity of flavors as some other vegetables, most notably tomatoes. Not that there aren't differences among varieties, but people don't come to fisticuffs over eggplant flavor the way they do defending their favorite tomato's reputation. Eggplant is one of those vegetables that is best when harvested young and mild.

It's worth trying different varieties, but in my opinion most eggplant varieties taste like . . . eggplant. As long as it's harvested before biological maturity, you will have done your part as the grower to maximize flavor potential for your eaters. This can be an advantage of greenhouse eggplant over field eggplant harvested when overmature.

Post-Harvest/Storage and Selling

Store harvested eggplant at temperatures of 54 to 59°F (12–15°C) and 80 percent relative humidity. Despite the thick skin, it's prone to dehydration at low humidity. The skin is thick enough to protect the fruit for direct marketing and short durations on store shelves.

When selling greenhouse eggplant, make sure to play up the fact that the fruit was picked before becoming overmature, without bitterness, seeds, or thick skin. Mini eggplant may need to be packaged and sold at a higher price by volume to make up for lower yields.

CHAPTER THIRTEEN

The Leafy Crops

Alongside the vining and fruiting crops we've just profiled, leafy crops – including lettuce, herbs, and many species of greens and microgreens – represent the other most popular group of greenhouse crops. Commercial greenhouse production is profitable because there is year-round demand for fresh greens, and the price is high enough to justify the expense of a hoophouse or greenhouse.

However, there are major differences between growing fruiting crops and growing greens. The light, heat, and fertility needs of greens crops are all lower than those of fruiting crops. Plants just don't need as much energy to make leaves as they do to grow the plant and crank fruit out at the same time. Most of the commonly grown leafy species (basil being a notable exception) can tolerate and even thrive at lower temperatures than the fruiting crops. So if high inputs of energy and fertility are not desired or possible, greens production offers a good alternative. On the other hand, most greens crops will not grow as well or taste as good at high temperatures, so growing greens crops may not be the best option for hot areas during the summer.

Leafy crops are more straightforward in some respects – for example, there is no grafting, trellising, or crop steering involved – so the information is likewise more concise. Though much of the introductory matter in this book applies equally to leafy and fruiting crops, the basics of growing leafy crops in protected culture require much less space. Which is not the same as saying that they're simpler. The complex interplay among variety, management decisions, weather conditions, pest and disease pressure, and a million other factors in the environment can lead to a lifetime of discovery with any one of these crops.

Maximizing Leafy-Crop Production

Because the vining/fruiting crops have such a long establishment period before they bear any fruit, maximizing fruiting-crop production in protected culture depends on maximizing the amount of time spent harvesting compared with time establishing the crop. Since the establishment period is fairly constant (most large tomatoes, for instance, are going to need sixty days or so to go from transplant to harvest regardless of variety), the only way to meaningfully increase the ratio of harvest time to establishment time is to increase the length of harvest by making the season longer. Another way to look at it is that varieties aren't going to get that much earlier, so to get more out of a season you have to harvest for a longer period of time.

Cropping cycles for greens are much shorter, so one of the main ways of increasing greens crop production is to grow the crop to maturity as quickly as possible and minimize the days to maturity. In this manner, you can cram as many crops as possible into a single season. For example, let's say you're growing a lettuce variety that takes thirty days to mature. If you can grow that variety two days faster, or find an earlier variety that shaves two days off the production time of each crop cycle, that would save twenty-four days in year-round production, which would give you almost one more crop per year.

Temperature and Shade

If you're trying to grow greens crops to harvest as quickly as possible, note that higher temperatures will result in faster growth. But this is where the maximum temperature needs to be taken into account. In addition to increasing the rate of growth, going above the maximum recommended temperature for leafy crops will also hasten the speed of bolting. So the maximum temperature should be respected as much as possible, since many greens develop off flavors when grown too hot, even if bolting doesn't occur.

In many areas in the summertime, it's a struggle to keep greens crops below the maximum temperature. When outdoor temperatures are above the maximum, it's almost impossible to keep temperatures within the ideal range in a structure. This is where shade cloth and whitewash can come in handy. When put on or over a structure, they can significantly reduce temperature increase due to solar gain.

There is a happy medium to strive for in using shade. In moderation, shading can actually make leaves grow bigger, as plants try to adapt to a lower-light environment by making bigger "solar panels." When light gets too low, however, plants just get leggy, which means overly long stems that don't usually result in marketable greens.

Unfortunately, there is no chart that can express the magic ratio among crop, temperature, amount of shade, and optimal production without legginess, especially since outdoor temperature is never a constant that can be planned for. The best strategy to determine the amount of shade to use in any area is to talk to neighbors and shade retailers to find out what has worked in the past in a given area. In very hot areas, there may be no amount of shade that will make lettuce and some other greens crop production economical in the summertime, which is why many growers in hot areas take the summer off instead of the winter.

Regardless of climate, growth rates will be slower in the winter. Even if a greenhouse is heated adequately, the winter days are shorter and offer less light, and whole-greenhouse lighting for production isn't usually economical to make up for the loss of light in winter. But the seasonal difference in growth rates can be used to help plan the size of a greens greenhouse or hoophouse. Let's say that you have a market or account that takes ten thousand heads of lettuce every week, year-round. You know this quantity requires two bays of a greenhouse to produce in the winter, though only one and a half to produce the same amount in the summer. You'd want to make sure to build at least enough production area to meet the wintertime demand. In summer, that space can be used for cover-cropping, another cash crop, or planting extra lettuce in case the weather gets too hot and some bolts.

Planting Density

While the crop cycles for vining/fruiting crops and greens crops may differ, one factor for maximizing productivity in protected culture is the same for both: high planting density. No matter what leafy crop you're growing, getting the most out of your protected growing space also depends on planting as densely as possible. As with any crop where you harvest the whole plant, the only ways to increase yield are by packing more plants into the space, or packing more plants into the same amount of time. The adaptations that allow greens to be planted densely are compact frame size and resistance to diseases related to high humidity and poor airflow.

One of the most important approaches to maximizing planting density is varietal, which is why it's important to try new varieties. Finding a variety that is dense and compact, that can produce the same weight of greens in a smaller space than a current variety, or that is disease-resistant and can be planted more densely, can make a big impact on the bottom line. Compact frame size means the ability to fit more plants into a smaller area, which is why most lettuce varieties bred for protected production are dense and compact.

A great example of a variety that does *not* fit this description is the old looseleaf lettuce variety Black Seeded Simpson. Once a standard in field production and reportedly around for 150 years or more, Black Seeded Simpson was popular because it quickly produced a large, fluffy head of light-green leaves. Tons of this variety are still grown, but not so much in greenhouses. The fact that it's quick is great, but since it takes up so much room, protected growers have come to favor more compact varieties that fit the same weight of lettuce into a much smaller space than Black Seeded Simpson. See table 13.1 to see how much more profitable protected space is if it's planted more densely.

As far as disease resistance goes, there is an inverse relationship between planting density and the amount of airflow between plants, so varieties that are more resistant to diseases of reduced airflow, like downy mildew, will do better in high-density plantings.

Hoophouse growers, or greenhouse growers who favor moderate heating, may be able to grow leafy crops effectively with only minimal heat during the winter when there isn't enough heat and light to grow fruiting crops. A cropping schedule with greens growing in the winter and fruiting crops during the rest of the year may provide year-round income for growers who don't want to heat enough to grow fruiting crops throughout the whole year. Another advantage of this schedule is that it will provide labor for workers year-round; it can be just as important to keep good workers as it is to hold on to markets by growing year-round.

Planting Density Scenarios

The best way to illustrate the importance of plant density is by applying several different days-to-maturity and planting density scenarios to the same space. Let's say you have a 4 × 100-foot (1.2 × 30 m) bed open and you're trying to decide what to plant in it. You can either plant full-sized heads of lettuce at 12-inch (30 cm) spacing, mini heads at 8-inch (20 cm) spacing, or mini heads at 6-inch (15 cm) spacing.

With five heads across the bed, you can get a total of five hundred heads out of a hypothetical bed at 12-inch (30 cm) spacing. You can see how, by decreasing the spacing to 8 inches (20 cm) in each direction, you can get 7 heads across the bed and more than double the number of total heads, up to 1,050. When you halve the amount of space between heads down to 6 inches (15 cm), you more than triple the number of heads you can plant, up to 1,800.

It gets really interesting when you add start adding a selling price into the equation. Let's say you can get $1.50 for your heads of lettuce, regardless of whether

Table 13.1. How Planting Lettuce More Densely Can Increase Revenue

Spacing	1' (30 cm)	8" (20 cm)	6" (15 cm)
Number of heads per bed	500	1,050	1,800
Value per planting @ $1.50/head	$750	$1,575	$2,700
Total yearly value with 10 crops/year	$7,500	$15,750	$27,000

they are full-sized heads grown at 12-inch (30 cm) spacing or smaller heads grown at tighter spacing. Your five hundred full-sized heads would bring in $750 per bed, whereas heads at 8-inch (20 cm) spacing would bring in just over twice as much at $1,575, and heads planted at 6 inches (15 cm) would bring in almost four times as much, at $2,700.

Even supposing that you got a premium for a full-sized head of lettuce, you would still make more growing mini heads and selling them for a lower price. If you got $2 for your full-sized lettuce heads and $1.50 for the minis, you would still only make $1,000 per bed at 12-inch (30 cm) spacing, versus $1,575 at 8-inch (20 cm) spacing for $1.50 and $2,700 at 6-inch (15 cm) spacing for $1.50.

Ultimately, what you grow has to fit with what your markets want to buy. But with such pronounced differences, it's worth finding out if your customers will buy smaller heads of lettuce. Also, it has been our experience selling at farmers markets that people don't always want to pay more for larger heads of lettuce. Big heads of looseleaf lettuce may be seen as a standard product not worth paying a premium for, even if they do weigh more. And small, dense heads of lettuce may be viewed as a fancy quality product worth paying a premium for.

When we apply the factors of planting density, price, and time to our equation is when the differences become really pronounced. Let's say you have a heated greenhouse and can get ten crops a year. That 4 × 100-foot

Figure 13.1. As demonstrated by this organic endive propagated in soil blocks, you can transplant the blocks by placing them on top of the soil and watering thoroughly with overhead irrigation. Having the base of the plant off the ground maximizes airflow despite close spacing. See the author's fingers in the picture for scale.

(1.2 × 30 m) bed planted with heads of lettuce 12 inches (30 cm) apart could bring in as much as $7,500 at low planting density. The medium 8-inch (20 cm) spacing could bring in over 50 percent more, at $15,750, and the 6-inch (15 cm) spacing would almost triple your money at $27,000 per bed.

Now, you can't expect to plant a big head of looseleaf lettuce at 6-inch (15 cm) spacing and get marketable results. Though there are some varieties that will grow smaller when crowded, other large-framed field varieties are prone to disease and rot due to the decreased airflow and increased humidity when planted at increased densities. We have definitely noticed that some varieties we can grow at 8-inch (20 cm) spacing tend to rot at a 6-inch spacing, depending on the season, which is why trialing is so important to find the varieties that grow the fastest and tolerate the highest planting densities.

One way to mitigate the effects of dense planting is to make sure not to transplant too deeply. This is especially useful for lettuce and the heading greens where, when transplanted deeply, the base of the head gets buried and traps a lot of moisture. This can lead to rot on the bottom of the head when it stays moist for extended periods of time, especially when ventilation is poor. This is a particular problem with lettuce in soils infected with sclerotinia, which leads to the appropriately named disease bottom rot. This rots the lettuce from the ground up and makes it unmarketable.

In Europe, shallow planting has been used for decades in greenhouses that use soil blocks to minimize bottom rot. At transplant time, holes are dibbled that are the same dimensions as the soil block. When the crops are transplanted, they are simply dropped into the hole and watered in to make sure the blocks don't dry out before the roots grow into the soil. This is a great reminder that transplants don't need to be buried very deeply, which can reduce both labor spent firming plants into the soil and losses to bottom rot.

Lettuce and Greens

Lettuce has long been one of the most popular protected culture crops. The year-round demand for fresh salads and the fact that it has limited shelf life means there will always be demand for fresh lettuce. Historically, greenhouse production has focused on butterhead lettuce, partially to differentiate it from the iceberg and looseleaf types grown in the field for shipping.

Greenhouse growers and consumers like butterheads because they tend to have very good eating quality. Butterheads still make up the largest group of lettuces bred specifically for protected production, though there are many other types now that are either bred for or adapted to protected growing.

The many species comprising the general term *greens* – such as kale, arugula, and mustards – are more recent arrivals on the protected cropping scene in North America. Since most greens do well growing at the same temperatures as lettuce, growers can plant mixed crops based on market demand. Demand for premium salads with ingredients besides lettuce has led to the cultivation of a wide variety of greens year-round, which soil growers have the option of direct sowing at high densities for baby leaf production.

Traditionally, greenhouse lettuce was grown during the winter in rotation with fruiting crops. Though this is still frequently done, there are now many growers who specialize in year-round lettuce and greens production.

Variety Selection Criteria

The most basic consideration when deciding what varieties to grow is whether you want to produce heads or salad mix. If salad mix is the finished product, one-cut lettuces will give the best yields of high-quality salad mix. Other than the one-cuts, there are many varieties of lettuce and greens that have been bred specifically for leaf shape at the baby stage.

There is no equivalent to one-cut lettuce for the other greens species yet, though vase-shaped brassica species that have all the leaves attached at one basal point, like bok choy, komatsuna, Tokyo Bekana, tatsoi, Yukina Savoy, and mizuna can be used much the same way. Entire heads can be harvested and the base removed during processing to release the individual leaves. Or loose greens can be harvested by cutting just above the base to release the individual leaves.

Bolting Resistance

Resistance to bolting is an important trait for all types of lettuce and greens. Bolting is when the plant starts forming a flower stalk in preparation for making seed. It is an indication that the plant has gone from a vegetative state, focusing on leaves, to generative growth, putting its energy into stem elongation and flower and seed formation.

Bolting is important because it's associated with a rapid decline in eating quality and marketability. As soon as bolting begins, lettuce usually starts becoming bitter. The stem also begins to elongate, which gives bolted lettuce its characteristic pine tree shape. If harvested early enough in the process, slightly bolted lettuce may still be marketable, but it should be tasted for quality. No one wants to buy a tall, pointy, bitter lettuce.

If there is any doubt, the way to tell if lettuce is bolting is to harvest a head and cut it in half lengthwise. In lettuce that has not bolted, the stalk (the point where all the leaves are attached) is short and disk-shaped. In lettuce that has begun to bolt, the stalk becomes elongated (see figure 13.13). At first this may not be noticeable, but eventually the elongation of the stalk beyond the lettuce's frame is what give it the distinctive pine tree shape.

Breeding puts lettuce under two distinct and conflicting selection pressures. On the one hand, we want varieties that will grow to a marketable size as quickly as possible. On the other, we want lettuce that will hold at marketable size for as long as possible, without sailing past maturity into bolting. Trying to have both traits at the same time – rapid maturity without rapid overmaturity – is job security for lettuce breeders.

Another part of the bolting puzzle is that weather conditions that stress the lettuce plant – especially

Ideal Lettuce and Greens Climate

OPTIMAL TWENTY-FOUR-HOUR AVERAGE TEMPERATURE

The most productive average temperature for growing lettuce is 63 to 65°F (17–18°C).

MIN/MAX TEMPERATURE

Try to avoid temperatures above 82°F (28°C), as this may result in bitterness and premature bolting in lettuce and greens. Drying winds can be more damaging to plants than cold temperatures, so in protection these crops may be able to go down below freezing for extended periods in a structure with protection from the wind and elements. How low and how long they can go depends on species, variety, and stage of maturity. Generally, younger plants are more resilient to cold temperatures than more mature ones.

CLIMATE

The fact that lettuce and greens can survive and even thrive at a much wider range of temperatures than fruiting crops means that growers have much more latitude in what kind of temperature regime they can use to grow the crop. Hoophouse growers in cold areas can keep leafy crops alive through some very cold temperatures, especially if a layer of row cover is used to cover the crop inside the structure as a second layer of protection.

One consideration for how hot to grow the crop might be whether or not your lettuce and greens production area is maxed out. If you have more area than you need to meet demand, you may be able to grow it more slowly and let it take longer in production. On the other hand, if your amount of growing area can barely keep up with demand, growing it as quickly as possible is a way to maximize the amount of production from a given area.

One way to think of temperature schemes for growing greens crops is that there is a cooler option and a warmer option for winter and for summer, and the temperature needs to be matched to the amount of light. Crops can be grown in the cooler range for the season if there is plenty of light, but growing them hot without a lot of light may result in them respiring more than they can photosynthesize, burning up most of the energy they produce and growing poorly.

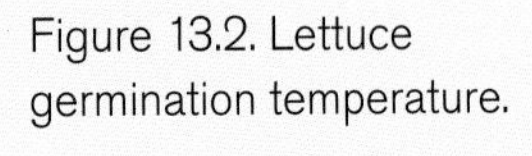

Figure 13.2. Lettuce germination temperature.

Figure 13.3. Lettuce and greens seedling growth temperature.

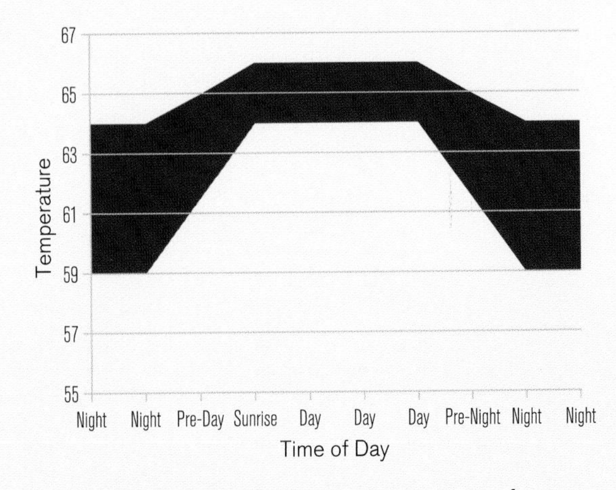

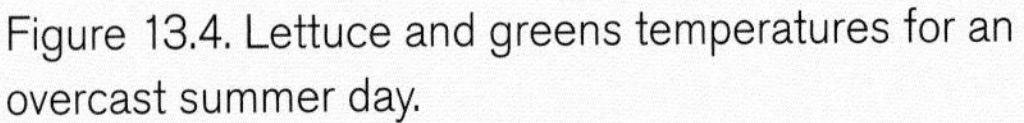

Figure 13.4. Lettuce and greens temperatures for an overcast summer day.

WINTER GROWING

On cloudy days or during the low light of winter, lettuce and cold-tolerant greens (like the brassicas) can be grown as cool as 50 to 54°F (10–12°C) during the day, and get as low as 43 to 47°F (6–8°C) at night. Many varieties of lettuce and brassicas can take some freezing temperatures as long as they are hardened off. Though temperatures below those above will not kill most greens crops, they will not result in much growth, either. You can maintain temperatures just above freezing to hold a crop until it's sold, using your structure as a cooler. But not much active growth will result at daytime temperatures below 50°F (10°C).

On sunny days in the winter, a temperature of 59 to 64°F (15–18°C) may be used, with night temperatures of 50 to 54°F (10–12°C).

SUMMER GROWING

On cloudy days in the summer, use a temperature of 64 to 66°F (18–19°C) during the day and 59 to 64°F (15–18°C) at night. On sunny days, use a temperature of 66 to 72°F (19–22°C) during the day and 59 to 64°F (15–18°C) at night.

Often the struggle in the summer is keeping the greenhouse cool enough. Shading that is applied to the covering, shade cloth, swamp coolers, or high-pressure fog may moderate the temperature. Use the most heat-tolerant and bolting- and tipburn-resistant varieties. If at all possible, keep temperatures below 82°F (28°C). Above this level, premature bolting and tipburn become much more likely.

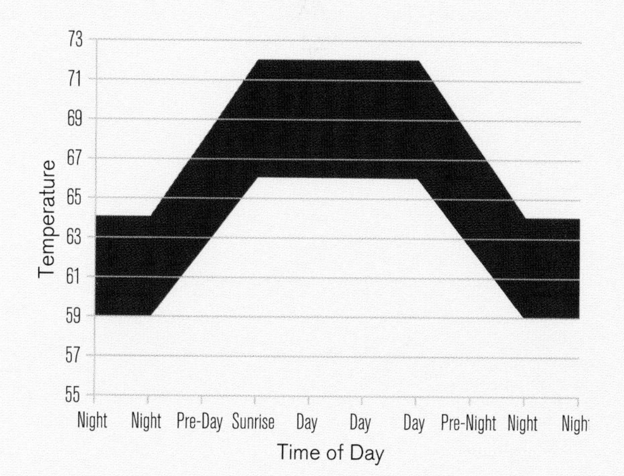

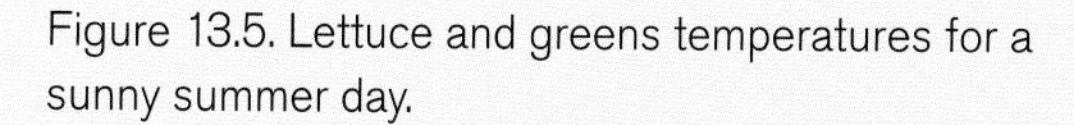
Figure 13.5. Lettuce and greens temperatures for a sunny summer day.

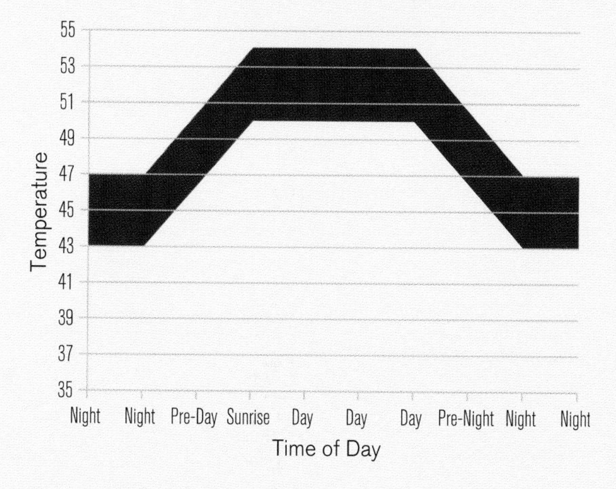

Figure 13.6. Lettuce and greens temperatures for an overcast winter day.

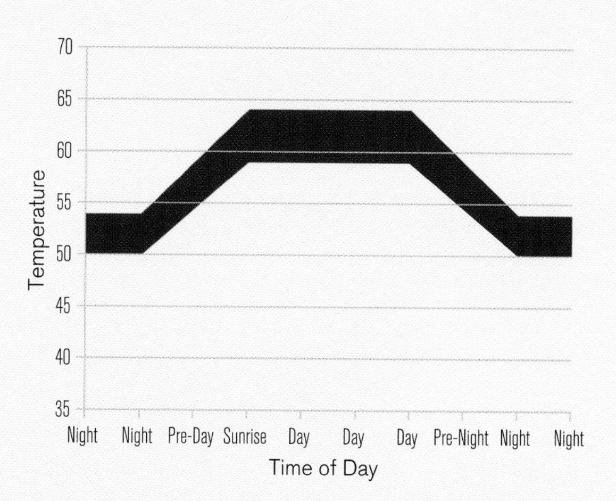

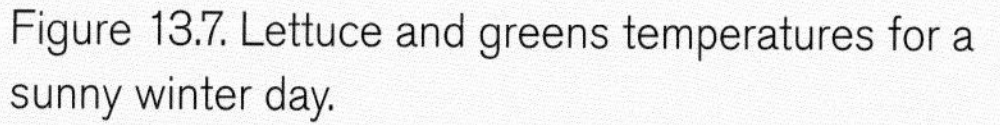
Figure 13.7. Lettuce and greens temperatures for a sunny winter day.

hotter-than-ideal or rapidly fluctuating—may initiate premature bolting, whether or not the variety has reached marketable size. As a grower, you want to find varieties that hold at marketable size and don't initiate bolting under adverse weather conditions for as long as possible.

Tipburn Resistance

Certain varieties are much more prone to this disorder than others. One variety may not show any symptoms under the same conditions where another is severely affected. Given that tipburn is one of the most significant disorders affecting protected culture lettuce production (see "Pests and Diseases" on page 211), select varieties that are resistant to tipburn under your conditions.

Coloration Under Low-Light Conditions

Being hit by ultraviolet light is what induces the red coloration in lettuce. Winter is the most difficult time to produce well-colored red lettuces in protected culture, because not only are light levels lower in the first place, but greenhouse coverings exclude some percentage of the remaining light. A common phenomenon is for varieties that attain a dark-red color in the summer to end up green or a light-brown color in winter protected production. Reduced UV radiation is usually the culprit. Some varieties are more sensitive than others to low light, meaning some red lettuces will color up better than others. Trial available varieties under winter conditions to see which ones develop the best color.

Selecting Varieties for Production of Heads

Head lettuces developed for salad mix production are a fairly recent development; head lettuce production where the head is sold intact has long been the standard for greenhouses, with butterhead lettuces leading the pack. This is partially due to the fact that many butterhead varieties are more resistant to tipburn, one of the great scourges of greenhouse lettuce production, than other types. Also, butterheads are compact, so you can fit a lot of lettuce in a little space, and butterheads are among the best for eating quality.

Though butterhead lettuce is still a standard of protected culture production, growers have found that a wide range of varieties work well. "Summer crisp" types can work particularly well in hotter-than-ideal growing conditions. There are a number of other species that make heads that also do well in protected culture, including endive, escarole, chicory, pak choi, tatsoi, and kale. Okay, so kale doesn't really head up, but it's not uncommon to see kale sold as a whole plant anywhere from teenage to full-sized.

One-Cut Lettuces for Production of Salad Mix

One-cut lettuce varieties have given protected culture growers some great new options for cut greens production. They are the first head lettuces bred to be grown to maturity for salad mix production. They are called one-cut because all the leaves on a head can be detached from the core with a single cut. One-cut-type lettuces have been developed by several companies, and are currently marketed under the brands Salanova, Eazyleaf, Multileaf, and possibly others.

The defining feature of one-cut lettuces is that all the leaves on the head are about the same size. This is in contrast with a normal head, which has larger leaves on the outside and smaller leaves in the interior. The leaves stay the size of salad mix, even at full maturity. The other important feature is that all the leaves are attached at a single point at the base of the plant, which is what allows them to be separated from the core with one cut. This is in contrast with the traditional way to produce salad mix by heavily seeding lettuce in multiple bands per bed, and cutting the immature leaves off the plant when it only has a few leaves. There are several disadvantages to producing salad mix with baby leaves for protected culture growers, including the fact that the returns aren't particularly high. A lot of greenhouse growing space and time is spent waiting for seeds to germinate and grow. When they can be harvested, baby leaves aren't heavy, and yields aren't particularly good.

In contrast, one-cut lettuce is usually transplanted to the greenhouse, so the slow-growing small-seedling stage is spent at high density in a propagation area. The

Figure 13.8. One-cut lettuces growing in a hydroponic greenhouse.

potential yield of one-cut lettuce is higher, and so is the quality and flavor. This is exciting since it's one of the few times when a production advantage is also an eating quality advantage. Production is higher because instead of getting a few small leaves once or twice off each plant, you can harvest a full-sized head that quickly breaks down into salad-mix-sized leaves. The crop is easier to weed as well, since you can get between individual plants, something that's not possible with densely seeded baby leaf greens.

Quality is higher for several reasons. For one thing, if you are able to get more than one cut off baby leaf lettuce, it usually comes with some ragged edges. Some of the leaves that were cut the first time continue to grow and are inevitably cut again on the second pass. These twice-cut leaves not only look ragged in a salad mix but are the first to break down and rot when bagged. This reduces the shelf life of the entire bagged mix to the lowest common denominator of these twice-cut leaves.

Shelf life is also better with one-cut lettuce because the harvested leaves are mature, instead of being at the baby stage. Mature lettuce leaves last longer before rotting. Another advantage of mature lettuce leaves is that they have better loft than baby leaves, which tend to be kind of flat when immature, even for frilly varieties. Loft is important because it fills up both bag and plate. Loftier mix is more attractive than a bunch of flat, wet leaves stuck to one another. It seems like a better deal to consumers, too, because a lofty bag looks bigger than a bag of flat leaves of the same weight. And if you have restaurant customers, in addition to the fact that it looks better, loftier lettuce mix will make a plate look fuller while using less lettuce.

The eating experience is better with mature leaves as well. Mature leaves have a more defined rib and crunch than baby leaves, and they also have better flavor. Because of these benefits, one-cut lettuces have taken off since their introduction with both field and greenhouse growers.

Lettuce from Seed to Sale

Once you've chosen your varieties, it's time to think about how to make the most of them. Below are lettuce-specific recommendations to take you from seed to sale.

Propagation

Fill flats with the desired medium and plant seeds on top of the cells or dibble holes. Saturate the medium before planting, and cover the seeds with a thin layer of fine to medium vermiculite. Lettuce seed may germinate better in the light, so don't cover it too thickly unless you're using pelleted seed.

Figure 13.9. Certified organic lettuce seedlings, planted in soil covered with a thin layer of vermiculite on the author's farm.

Thermal Dormancy and Primed Seed

This is not a problem with the other species of greens, but one challenge to growing lettuce through the summer is thermal dormancy: the tendency for lettuce seeds not to germinate under high temperatures. Thermal dormancy is an evolutionary adaptation so that lettuce in the wild would not germinate in conditions hotter than it could grow in. Lettuce germination percentage will go down sharply when germination is attempted at temperatures hotter than those advised in this chapter. Thermal dormancy can be induced if seed storage conditions are too hot, which is why some growers store their lettuce seed in a refrigerator or freezer.

Primed seed is one way around thermal dormancy and other challenges to germinating lettuce. Priming is a specialized treatment that begins the germination process and then stops it, so the seed can be held in a partially germinated state. It can be done in a manner that's compliant with organic certification, so there is such a thing as organic, primed seed (most of the organic, pelleted lettuce seed, for example, is primed; see below). Priming may increase germination percentage and uniformity, especially when you're germinating under less-than-ideal conditions. The process was originally developed to increase uniformity in open-field lettuce, but it can be equally useful for germinating lettuce seeds in greenhouses that may be too warm.

Primed seed is frequently, but not always, pelleted. Priming also overcomes any inhibition lettuce seed may have to germinating in the dark so that it can be pelleted. Pelleting makes lettuce seed easier to plant by hand or mechanical means. Lettuce seed is so small and thin that it can be hard to handle. Pelleted lettuce seed is almost always primed, since the coating forming the pellet blocks light from reaching the seed, which could interfere with germination. Other greens species are not primed and pelleted as frequently as lettuce, but sometimes very small-seeded crops such as arugula are pelleted just to make the seed easier to handle.

The downside of priming is that the seed does not last as long as unprimed seed. The embryo cannot exist in "suspended animation" for as long as the dormant state of unprimed seed. It is recommended that you hold primed seed for a year or less. So don't plan on buying more than a year's supply of primed varieties.

Figure 13.10. Certified organic Swiss chard in a high tunnel in France.

Germination

Lettuce germinates best at a constant day/night temperature of 61 to 65°F (16–18°C). Above 68°F (20°C), germination may start to decline for sensitive varieties. Above 77°F (25°C), germination may be severely affected. Other greens crops, especially the brassicas, are not as particular, and these temperatures may work well for mixed plantings of lettuce and other greens species. It is worth finding the optimal germination temperature if you're seeding a lot of any one species of greens.

Seedlings

A good temperature for growing seedlings of lettuce and many other greens is a constant day/night temperature of 59 to 64°F (15–18°C). Greens crops do not need a break from light the way fruiting crops do, so if you have lights a twenty-four-hour photoperiod at these temperatures will make the seedlings grow faster.

Figure 13.11. Certified organic spinach in a high tunnel. Planting density is maximized by having only one pathway.

Figure 13.12. This partially harvested organic celery shows how such densely planted crops need to be harvested in blocks.

Baby Leaf Lettuce and Greens Production

There are many varieties of lettuce and other species, including herbs, that have been bred specifically or are suitable for cutting at baby size for salad mix production. The main considerations are flavor, leaf shape, loft, and weight. The best mixes have all of these.

Producing economical amounts of baby leaf greens is dependent on planting at high densities to get as many leaves from a given area as possible. The most efficient and accurate way to accomplish high-density seedings is by using a multirow seeder, such as a four-row pinpoint or six-row seeder.

These are fairly easy to use because they are designed for baby leaf greens, planting rows 2.25 inches (6 cm) apart, which is a good spacing for most species. Simply plant your desired bed size solidly at this spacing and come back and cut at your desired size.

Crop Cycles and Timing

Because lettuce and greens crop cycles are so much shorter than the fruiting crops, and profitability is determined by cramming as many plants as possible into a bed and as many crop cycles as possible into a season, there are no set seasons to lettuce production. The main questions are whether to produce year-round, and at what temperature to grow the crop.

Because lettuce and greens grow at a much wider range of temperatures than the fruiting crops, growers can grow them at fast, medium, or slow temperatures depending on outside temperatures, amount of sunlight, price of fuel, and amount of demand in relation to production space. See "Ideal Lettuce and Greens Climate" on page 204, for how to plan and manage these options.

Fertility

Lettuce and greens have similar fertility needs. Soil growers should get an SME soil test and amend accordingly. Romaine usually has higher nitrogen needs than other types of lettuce. However, keep in mind that too much nitrogen at any stage may result in excessively lush growth that is more susceptible to many types of pests and pathogens.

Figure 13.13. The cut-open head on the right shows the elongating core typical of bolting lettuce.

Labor

The amount of labor required to grow and harvest a lettuce or greens crop varies widely. At the high end of the spectrum is soil-grown baby leaf lettuce or greens, which may need to be cut by hand. This takes a long time, and the ergonomics are poor. There are a variety of greens harvesters ranging in size from handheld to small tractor size, which may speed up the process. This is yet another way in which one-cut lettuces can help with salad mix production. Alas, the equivalent for greens has not yet been bred, but no doubt someone is working on it. It is much faster to harvest heads of lettuce and sell them whole, or to process them into their individual leaves in the comfort of a packing area, than to cut handfuls of leaves while bent over in the greenhouse.

Flavor

Lettuce comes in a range of flavors and textures, which give each variety its particular culinary appeal. Some varieties are very mild and mostly eaten for their crunch, while others are prized for their sweetness or buttery mouthfeel. There is a much wider range of flavors in the other species of greens, from the lettuce-like Tokyo Bekana to the peppery bite of arugula.

Regardless of which species you grow, the best flavor will be produced by harvesting healthy plants at the proper stage of maturity, before they begin to go to seed. Plants that are unhealthy or under stress may produce bitter compounds or other off flavors. Bolting tends to make lettuce and greens bitter.

Pests and Diseases

See appendix B for more information on pests.

STEM ROT

Stem rot is most likely to occur in dense lettuce and greens plantings where there isn't sufficient airflow, and is pretty much what it sounds like: Lesions form with the potential to rot the portion of the plant between the soil and the first few leaves. Stem rot can be reduced by increasing ventilation, planting more shallowly, and making sure irrigation water is free of pathogens.

BOTTOM ROT

Bottom rot is caused by the pathogen sclerotinia, which thrives in the moist area between the soil and the base of the plant. Bottom rot may be reduced by planting seedlings less deeply so there is more airflow between the base of the plant and the ground. Be sure not to bury the base of the plant.

LETTUCE MOSAIC VIRUS (LMV)

This is the most significant lettuce virus, and results in mottling on young leaves. It can be seedborne and spread by aphids. Many varieties are indexed for LMV, which means that they test and confirm a certain number of seeds out of each lot to be free of the virus. This is not a guarantee that the entire lot is LMV-free, but it is as close as you're going to get. There are also some resistant varieties.

Harvest

Whenever possible, harvest leafy crops early in the day or when temperatures are as low as possible. Crops will have better shelf life when they are not transpiring quickly at harvest. Remove the field heat by immersing in cold water to cool down to storage temperatures quickly. If lettuce or greens are frozen, wait until they have thawed to harvest. Leaves that are harvested when frozen will collapse when they warm up, even if the species can withstand freezing weather.

Post-Harvest/Storage

One way some lettuce and greens growers who don't quite grow year-round bridge the sales gap is by storing leafy greens. Most varieties of lettuce and greens store well at very high humidity of roughly 95 percent, just above freezing at 33°F (1°C). Under these conditions lettuce and greens may last two to three weeks, though it is ideal to sell them well before this so they have some shelf life in the store or your customer's refrigerator.

Mature leafy crops that were exposed to some cold temperatures and stored in ideal conditions may be salable beyond a three-week period. Greens products harvested when immature usually will not store for as long as mature leaves. One way to extend the storage life of salad mix for growers is to harvest heads of multileaf lettuce, put them into storage whole, and only process them into salad mix as needed to meet market demand. In this way the normal storage limitations of baby greens can be exceeded.

Selling

Some head lettuce, particularly that which is packed with its roots intact, is sold in clamshells, sleeves, or bags to keep it clean and labeled in transit. Most of that which isn't bagged is packed in boxes to the specifications of the buyer. Most greens mix is bagged, or put in plastic boxes for larger retail amounts. Wholesale amounts of greens mix is bagged and put into larger cardboard boxes.

Basil and Other Herbs

The most widely produced herb in protected culture is basil. There is so much more basil produced than any other herb that it is the only one with its own system, as far as best practices and varieties developed specifically for protected culture go. The fact that it's so widely used and very perishable means that there is high, year-round demand for fresh product. I know some greenhouse tomato growers make more money on basil than on their tomatoes and plant as much basil as the market will take.

Other fast-growing culinary herbs such as cilantro and dill may be profitable because their short time to maturity means the space can be turned over many times in a season. Many of the other perennial and slow-growing herbs may be harder to make a profit on in protected culture, since they cannot be turned over as quickly.

One of the benefits of growing some woody herbs in hoophouses is that they may be overwintered in some areas where they otherwise would not survive. For example, the most cold-hardy rosemary variety, Arp, will overwinter in unheated hoophouses in many areas of the northern tier of the United States and southern Canada. In addition to allowing early cut herb sales, crops that can be vegetatively propagated have the additional value

Ideal Basil Climate

OPTIMAL TWENTY-FOUR-HOUR AVERAGE TEMPERATURE

The most productive average temperature for growing basil is 75 to 77°F (24–25°C).

MIN/MAX TEMPERATURE

Basil is very sensitive to cold temperatures, and grows best between 70 and 80°F (21–27°C). Basil can grow at higher temperatures, but this may cause the plant to go to seed more quickly. It's undesirable for basil to bolt, because the plant starts to put its energy into flower and seed production instead of leaves.

Depending on how you are selling it, it may be possible to prolong vegetative growth temporarily once flower production has started by removing the flowers, which are produced at the tips of the branches. But once the plant has gone reproductive, it's never going to put as much energy into leaves as when it was young. If you are doing cut and come again (CACA) harvesting, it's a good idea to have multiple successions planted so when one goes to flower, you can start harvesting a new planting.

CLIMATE

Basil seed germinates between 70 and 75°F (21–24°C). Basil grows well between 70 and 80°F (21–27°C). It can survive temperatures up to 100°F (37°C), though it will require more water and will not grow as well as the upper limit is reached. It does not need much difference from day temperatures, with a night temperature of 65 to 70°F (18–21°C) providing good growth. Temperatures below 50°F (10°C) may cause leaves to turn black and die, making it unsalable.

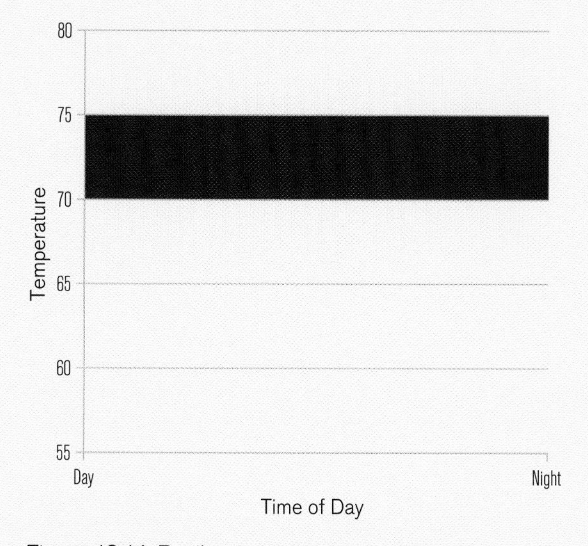

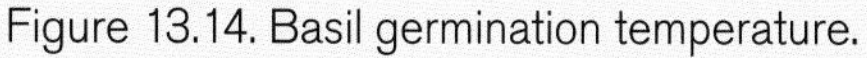
Figure 13.14. Basil germination temperature.

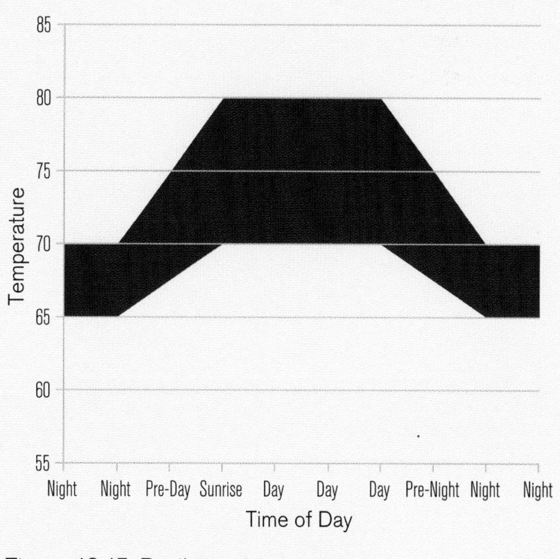

Figure 13.15. Basil growing temperature.

of serving as mother plants, so early cuttings can be produced for seedling sales or farm use.

Many species of herbs in protected culture play nicely with the other leafy crops, and can be interspersed with lettuce and greens production in response to demand without special treatment. On the other hand, if you are growing a lot of any one herb, it might be worth devoting a separate production area to that crop.

For instance, if you want to devote half of your production space to lettuce and half to basil, you'd need to consider that basil likes a warmer environment than lettuce does. When it comes time to set the climate in your basil and lettuce production area, you would have three options: set the climate for lettuce, set the climate for basil, or compromise and make neither crop completely happy. Of course in the real world compromises like these happen all the time, but on the other hand it means that some part of your greenhouse is not running at optimal production. This would be particularly pronounced in the winter. You would waste a lot of fuel keeping the mixed lettuce/basil greenhouse hot enough for basil. A simple solution would be to install a plastic curtain to segregate the climate in one bay from the other(s). This would necessitate installing a separate set of climate sensors in each bay, but it would allow you to control them independently, save fuel, and optimize growth.

Varieties

Since there is more greenhouse basil produced than any other herb, there is probably more breeding for basil in protected culture than for all the other herbs put together. For many herbs that are not routinely produced in protected culture, there may not be any varieties developed specifically for greenhouse production. It's worth inquiring with seed companies that carry greenhouse varieties to see if they have any available. Even if it's not a regular catalog item, they may have access to herb varieties for protected culture.

Whether it's specifically developed for the greenhouse or not, the traits you should look for in greenhouse herbs are fast growth, an open plant habit for good airflow, the ability to grow well under reduced-light conditions, and a compact growth habit. For varieties with red or purple foliage, it's important that they develop good color under low-light conditions. This is especially important for winter production when light levels are most restricted. Different varieties have different thresholds for developing acceptable coloration.

Another consideration is whether there are any varieties that have resistance to the diseases you may face in protected culture. The main resistances that are available at this time are to fusarium and downy mildew in basil. Unfortunately, high resistance has not yet been developed for downy mildew, which is becoming a huge problem in basil production.

Basil from Seed to Sale

Once you've chosen your varieties, it's time to think about how to make the most of them. Below are basil-specific recommendations to take you from seed to sale.

Propagation

Plant seeds into your desired medium and cover with a thin layer of vermiculite. Just as with the other leafy crops, going straight from propagation to production will save labor, since there is no potting-up stage in between. On the other hand, using a nursery stage is more conservative of space: The plants can be propagated in smaller cells since they will get potted up.

Direct seeding basil can be accomplished with a precision seeder such as the four-row or six-row seeder. Direct seeding works better for crops that are going to be harvested small and bunched, because it allows for very close plant spacing and efficient use of space. Transplanting works better for plants that are CACA-harvested, or sold at a larger plant size, because it allows each plant more space.

Planting Density

Full-sized basil may be planted anywhere from twelve plants/m^2 in the wintertime to twenty-four plants/m^2 in the summer. Vary planting density corresponding to the amount of light received. To speed up the production cycle, use the same planting density and multiple seeds/plants per cell. Soil growers also have the option of planting densely, and pulling or cutting plants and bunching.

Crop Cycles and Timing

Herbs (including basil when the whole plant is harvested) fit the short-cycle production model, where the amount of profitability is determined by how many crops you can get out of a particular production spot or amount of greenhouse space. The exceptions to this model would be basil or sage, or anything else that is CACA-harvested. This may only stretch out the production cycle slightly, however, as most culinary herbs eventually go to seed or decline in productivity after being harvested for a while.

Labor

Requirements are similar to those for the other leafy greens, and vary depending on the system used to maintain the plants. See "Labor" in "Lettuce and Greens" on page 211.

Pruning

Herbs are not pruned in the traditional sense. If you're harvesting by cutting repeatedly, most herbs including basil will form new branches from the lateral buds below where each cut is made.

Yields will be increased by removing flowers as they form. Otherwise the plant will put its energy into flower and seed production. It's a good idea to have another succession ready when basil goes to flower, because yields will never be as high as when the plant is focusing on vegetative growth.

Pests and Diseases

See the descriptions of fusarium and downy mildew in appendix B. Use resistant varieties as much as possible. Downy mildew in basil is a recent phenomenon in North America, so at this point no varieties are available that are completely resistant. This has become a huge problem for

Figure 13.16. Basil sharing space with cucumbers in one of the author's hoophouses.

field and protected basil growers alike, so find out what varieties have the highest resistance at the time of planting.

Flavor

Flavor varies widely based on variety. The most widely used types of basil are variations on Genovese, which is the standard for flavor used in Italian cooking. Thai basil has smaller leaves and an anise-like flavor. There are a number of other variants that may have citrus or other flavors, which there may be specialized markets for.

Harvest

Basil may be sold as living product, still growing in a pot. Or immature plants may be cut and bunched, or pulled from the soil and bunched, at almost any size. If plants are pulled from the soil with roots intact, as much soil as possible should be washed off the roots to give the bunches a clean appearance, and to keep the soil on the farm. For fresh-cut leaf, plant tips may be cut, or entire baby plants may be cut and sold loose or bunched.

Post-Harvest/Storage

Most culinary herbs store well just above freezing, at 33°F (1°C). Basil requires a much warmer storage temperature and is best stored at 50°F (10°C). If it's exposed to lower temperatures it may turn black, which will make it unsalable. All herbs should be bagged or stored at high humidity to minimize water loss from the leaves. Many herbs are sensitive to ethylene and should not be stored with ethylene-emitting produce.

Marketing and Selling

Most cut herbs are sold bunched, bagged, or in small rectangular clamshells to keep them clean and retain moisture. If your herbs are packaged, make sure your sticker is on the finished product to let customers know where it was grown and any special attributes such as being organic or pesticide-free.

At farmers markets, herbs—especially basil—are sometimes sold loose by the pound, though we have found that people may buy such small amounts as to make it not worth selling, which is why I recommend bunching or pre-packaging so you can limit the size of the minimum sale.

Though it isn't usually economical to grow protected herb crops with the intention of drying them, drying can be a good backup plan, though you will get a much higher price by selling fresh-cut or living/potted herbs.

Herbs grown for drying are usually produced in the field where production costs are lower. Drying can be used to prevent herbs from going to waste when they come back from a farmers market, or in other retail situations where they have only been on display for a short while. We use this strategy to ensure we have a big herb display at farmers markets without a lot of waste. We harvest more herbs than we think we can sell, to make an abundant-looking display, and whatever comes back from market goes straight into the dehydrator. Herbs that are at the end of their shelf life will not make a quality dried product. But as long as they are still fresh, they can be put into a dehydrator and garbled (crumbled) when dry for later use or sale.

We dry and garble all the herbs that come back over the course of the season into bulk containers. At the end of the season, when the herbs in our hoophouses have died or stopped growing for the winter, we package up the dried herbs. In this manner we are able to stretch the season on herbs without adding heat, which probably wouldn't pay for itself except for basil. Packaged dried herbs make good holiday gifts so there is a strong market for them after the fresh herb season is over. We did have to get a license for handling the dried herbs, and regulations vary by state or province, so check with local officials before retailing these.

Microgreens

Microgreens (micros, for short) are simply the seedlings of any edible-leaved crop, grown anywhere from the cotyledon to three-true-leaf stage. They are baby greens taken to an extreme. Micros are very similar to sprouts, with some important differences. The term *sprout* usually refers to a seedling that is eaten whole, roots and all. Microgreens are grown slightly longer than sprouts, and are cut from the root. Microgreens are grown in a medium, whereas sprouts are not grown in anything except water. And sprouts are usually not grown for long enough to develop true leaves. Since they have true leaves, microgreens need light to grow properly, unlike sprouts.

The size and age of harvestable microgreens is determined by the preference of the end user. If a chef wants very small microgreens, they can be cut at the cotyledon stage. Most species of micros look more similar to one another at the cotyledon stage than they do later, since many of the differences between mature leaf shapes and colors are more apparent when the true leaves emerge. If customers want a larger micro with more of the character of the mature plant, they can request greens that have been allowed to grow longer and develop true leaves.

Microgreens are usually used as a garnish or accent, rather than as the bulk of a meal. They're most commonly used in restaurants to accent a plate of food. Some microgreens species are used simply to add visual interest, by adding interesting color, leaf shape, and texture to a meal. If chefs are looking only to augment the visuals of the meal, some micro species can be very mild while still providing an interesting leaf shape. Some species are used as a flavoring, providing a surprisingly large punch of fresh flavor. Basil, onion, and cilantro are examples of species that are used as micros for flavor that is sometimes slightly different than the flavor of the mature plant.

Microgreens can be made from any species that has edible leaves. See "Varieties," below, for more information. Growers are trying new species and varieties all the time in order to find the hot new microgreen.

Microgreens are the most unlike all the other crops featured in this book. For one thing, their crop cycle is much shorter since they're allowed to grow only barely past the germination stage. Microgreens' price per pound is also higher than almost anything else you can grow legally, though your seed cost will also be proportionally higher since the crop cycles from seed to harvest so quickly. It's not unusual for micros to fetch a retail price of $30 to $40 a pound. Another notable difference is that the retail market is very limited, since the main end user of microgreens is restaurants, though there are some microgreens sold retail at farmers markets and grocery stores. The mainstreaming of micros into peoples' diets at home would represent a potentially huge expansion of the market.

Yet another distinction is that microgreens are the only crop in this book that is exclusive to greenhouse production. Technically you could grow microgreens outside, but the barriers are great. The seeds and propagation medium are vulnerable to being blown around by the wind, and overwatered if it rains. And even if it never, ever rains where you live, the large amounts of seeds and young plants used in microgreens production tend to attract birds and rodents. Beside the fact that the predation would disrupt production, the droppings they leave behind would be a huge problem in a product intended to be consumed raw.

The outlier status of this crop presents a number of interesting dynamics for growers. Because of microgreens' uniquely short time to harvest, some growers first try microgreens by growing them at a time of year when they have unused propagation or production space. This can be a low-investment way for growers to test the market and profitability of the crop without investing in production space specifically for microgreens. For example, some growers may have propagation greenhouses that are only used for part of the year, and sit unused the rest of the time. Since microgreens can be produced on benches or the floor, growers can usually produce them with whatever space is available and decide if they are profitable enough to justify their own production space—because if the market proves to be lucrative, production may need to be year-round in order to hang on to customers.

Varieties

A wide variety of vegetables and herbs are used for microgreens production. A quick way to find varieties suitable for micro production is to look at the assortments offered specifically for that purpose by seed companies. Many catalogs offer both single varieties and blends of seeds for micro production that are selected to have a similar growth rate and make for a nice mix of colors and flavors. This is an easy place to start if you're new to microgreens,

Ideal Microgreens Climate

OPTIMAL TWENTY-FOUR-HOUR TEMPERATURE

Microgreens can be produced with many different species. If you're producing a lot of one type, it's worthwhile to look up the specific germination conditions and use those. For growers who are producing mixed micros, most of the species commonly used germinate well with soil/medium temperatures between 73 and 75°F (23–24°C).

Once the seeds have germinated, most species grow well with an air temperature between 70 and 75°F (21–24°C).

If the microgreens are being produced in a greenhouse or hoophouse with another crop, it's probably more important to tailor the climate to the other crop. Since microgreens don't go through an entire life cycle to maturity like most other vegetable crops, less-than-ideal conditions will probably have less of an impact on microgreens. When doing crop planning, keep in mind that cooler temperatures will produce slower growth and a longer time to harvest.

because they tend to include some of the most common and easiest-to-grow varieties. A pre-blended mix can also be used as the base of your own custom mix, by starting with a blend someone else has designed and adding your own custom varieties. Microgreens varieties are trialed mainly for days to maturity, appearance, and flavor at the micro stage.

Seed cost is another consideration. Because crop turnover is so rapid, and ratio of plant matter harvested to seed is so low compared with mature vegetable crops, you go through a lot of seed producing microgreens. There are some varieties of seed that might make a lovely micro that aren't sold for such because the seed cost would be prohibitively expensive.

One thing that could potentially be used to set your microgreens selection apart from other growers is using off-the-beaten-path varieties, those that are uncommon in microgreens production. Though the microgreens listing of a seed catalog is the best place to start looking, you can be sure that other microgreens growers have already tried those varieties as well. The trick to really customizing your mix is finding underused varieties that are affordable enough to use for micros, of which there is a consistent supply of seed. Though there may sometimes be cut-rate seed on the market due to overproduction or oversupply, consistency is the key to keeping your microgreens customers. You don't want to get customers hooked on some fantastic unique micro variety, only to have it drop out of the mix later.

The varieties used for microgreens are usually open-pollinated (OP), since they are cheaper to produce and stick around longer than the hybrids, which turn over more frequently than OPs. A very simple approach to developing your own mix would be to use a bunch of brassicas for volume, and some garnet-red amaranth for red color. Green and red brassicas, along with some amaranth for bright-red color, are some of the workhorses of the microgreens world.

There are so many species that are suitable for microgreens, and the end user tends to have such specific uses in mind, that it's wise to find out the customer's goals. Unless you have access to a wholesale market that will buy a standard microgreens mix, this is where getting to know your customer may pay off. If you have restaurants that use microgreens in your area, and you are willing to make deliveries to individual customers, microgreens may be a way to tap into a specialized low-volume/high-priced market. If you're trying to develop business with restaurants or other specialized customers, microgreens might be a way to get a foot in the door. If you can get them to start buying a signature microgreens product from you, they may start ordering other things you have available.

Selection Criteria

Whether the customer knows it or not, the three main things micros have to offer are color, flavor, and loft. Some varieties are only good for one of these features;

other varieties have more than one. Since microgreens are most often used as a garnish, the end user is usually thinking about how good your microgreens will make their food look on the plate.

Micros can be used to add a certain color to a dish. For example, chefs may use a bright-red microgreen to add visual interest to a plate that is otherwise monotone. Cooks sometimes use micros to add a certain flavor to the finished dish. Many micros offer a softer, more subtle version of the flavor found in the mature crop. It may be a good thing that not all micros are flavored as strongly as the finished product, because blends can be made to highlight certain flavors, with other varieties mixed in just for looks.

The final factor that has to be considered in micro mixes is the how three-dimensional the leaves are. Some chefs want micros that will add texture to a dish, whereas others just want to add a pop of color, like vegetable confetti. Varieties with very flat leaves, like most of the radishes, or micros that are harvested younger, tend to take up less space and make a denser mix. Other varieties with deeply cut leaf shapes, like some of the frilly mustards, can add a lot of loft to the mix and texture to the plate.

One production decision that has to be made is whether you want to plant a mix of varieties, or grow individual varieties and mix them after harvesting. The advantage of planting a mix is convenience: Just cut and they're ready to go. The disadvantage to pre-mixing microgreens seed, however, is that not every variety mixes well with every other variety. The main problem is growth rates. If you plant a very fast-growing micro (like a radish) with a slow-growing micro (like amaranth), the amaranth isn't going to look like anything by the time the radish is ready to harvest. So the advantage of mixing micros after harvest is that you can include any species you want. The disadvantage is that the planning is more complicated and it takes more time to cut the plants and then mix them together.

Commonly Used Microgreens Species

Most species that have become popular as microgreens lend an interesting leaf shape, color, or flavor to the finished product. Varieties that produce a color other than green are particularly sought after for making a mix with contrast, since most leaves are green. Many vegetables don't have strongly developed flavor at the micro stage. Finding something not on this list that makes a good micro would be a way to set your mix apart.

Amaranth, Red: One of the brightest reds you can add to a micro mix; not a lot of flavor.

Arugula: Adds a peppery, slightly spicy flavor.

Beets: Adds loft and color. Beet micros have brightly colored stems that match the color of the mature beet variety. Varieties like Bulls Blood also add dark-red leaves.

Brassicas: Many of the brassica species make good microgreens, including Chinese cabbage, collards, and various Asian greens. They tend to be fast growers that impart a mild brassica/broccoli flavor. Species of particular importance will be covered individually.

Cabbage: Red cabbage in particular is popular for the color that the purple-veined green leaves add to a mix.

Chard: Very similar to beet micros, chard adds some loft and the vein color of the mature plant.

Cress: Depending on the variety, combines an attractive parsley-like leaf shape with a slightly spicy flavor.

Herbs: Many herbs are used to lend the flavor of the mature plant, which tends to be milder in the microgreen. Some herbs that work well are basil, chervil, cilantro, cutting celery, dill, fennel, lemon balm, parsley, salad burnet, saltwort, shiso, and sorrel. The colored varieties of basil, fennel, shiso, and sorrel can also give a nice splash of color.

Kale: Many varieties are used for various leaf shapes, which are miniatures of the mature plant. Imparts a mild brassica/broccoli flavor. Red Russian is popular for its frilly leaves with purple edges and stems.

Kohlrabi, Purple: Produces micros with dark-purplish-green leaves and purple stems.

Komatsuna: Particularly important for the red varieties, which produce dark-red leaves.

Magenta Spreen: The common name for *Chenopodium giganteum*, a close relative of quinoa; it's mainly used as a microgreen because the leaves are tinged with hot pink, which stands out in a mix.

Mizuna: An Asian mustard, mizuna is important especially for its purple varieties, which add purple leaves and stems to the mix and a mild mustard green flavor.

Mustards: Among the most widely used microgreens. A wide variety of leaf shapes and colors are available. Some of the most important varieties include Garnet Giant, Red Rain, and Scarlet Frills. Most varieties have a spicy mustard flavor, so if you're looking for a mild mix, it's important to balance the mustards with some milder species.

Orach: Red orach is notable for being one of the darkest-purple/red microgreens available, with both the stems and leaves heavily colored.

Pak Choi: Red varieties can lend a nice splash of color to the mix.

Purslane: Important for the red varieties, which have bright-pink stems. The leaves and stems are very succulent, adding a nice crunch.

Radishes: Many radish varieties are popular microgreens. The large seeds quickly grow into a large seedling with good loft. Also, different varieties provide a wide variety of leaf and stem colors, from purple to white and everything in between. Most varieties have a mild radish flavor.

Scallions and Onions: Though they do not yield a lot quickly due to the thin, slow-growing leaves, members of the onion family are important in some mixes for the onion flavor they impart.

Microgreens from Seed to Sale

Once you've chosen your varieties, it's time to think about how to make the most of them. Below are microgreens-specific recommendations to take you from seed to sale.

Crop Cycles and Timing

Different varieties of microgreens have very different characteristics when it comes to days to maturity and volume produced. Most herbs are fairly slow growing at two to four weeks to harvest, whereas some fast-growing vegetables are ready in as little as one or two weeks. There are other slow-growing vegetables that take two to four weeks to produce. If you do sow a mix of micro seeds, make sure the varieties have comparable days to maturity. You don't want one variety approaching over-maturity as another one is just showing true leaves.

Use succession planting to ensure a continuous supply of microgreens. You'll need to plant new trays as frequently as trays are harvested, taking into account faster or slower growth depending on the season. Microgreens take the approach of cramming as many short cycles into a given year to an extreme. In theory you could have twenty-five or more cycles of fast-growing ten- to fourteen-day microgreens in a given greenhouse space in a year.

Note that while microgreens can be produced year-round in a heated greenhouse, production cycles will be longer if temperatures are lower in the winter. Microgreens may also be taller in the winter than under summer light levels. The other issue to be aware of with winter production is that lower light levels may not produce as dark a red leaf color.

Propagation

Microgreens are essentially always in propagation since they are sold before most other greens crops are

Figure 13.17. This is a tray of microgreens being grown on fiber substrate instead of soil.

Table 13.2. Microgreens Seeding	
Seeds per Pound	Amount of Seed to Use per 1020 Flat
100,000	25 g
100,000–200,000	20 g
200,000–300,000	15 g
300,000–400,000	10 g
500,000–750,000	5 g
750,000–1 million	3 g

Note: The larger the seed, the fewer the seeds per pound, and the more seed you need to use. The larger you want to grow the microgreens, the less seed you need to use. Use this as a starting point to find your own ideal seeding rate.

Figure 13.18. The larger seed on the right required 20 grams to cover the tray; the smaller seed on the left only needed 15 grams.

even transplanted. Most micros are only grown to somewhere between the one- and three-leaf stage. One of the tricks of microgreens production is to develop a seeding density and timing that will produce the desired product. A lower seeding density may be needed for the same variety if it's to be harvested at a larger size than at an earlier stage, since it will take up more space in the tray when harvested later. It's up to you to find your ideal combination of variety, seeding density, and days to maturity, but most microgreens can be planted with a spacing of 0.25 to 0.5 inch (6–12 mm) in all directions. An easier way to think about spacing might be by weight; see table 13.2 for a guide to planting based on seed size.

Micros can be planted in a wide variety of mediums. Trays full of potting soil are one option. Some growers use burlap or various other fabrics. There are some specialty fabrics specifically for growing microgreens, which are made from organic materials that can be composted. Almost anything can be used that is clean, that's nontoxic, and that will anchor the plant roots until harvest.

Make sure not to use treated seed for microgreens production, as the finished product is so close to the seed stage that fungicide or insecticide will linger on it.

Whether or not to fertilize is one of the big questions in microgreens production. It is possible to produce micros with a wide range of fertility. If fertilizer is applied through irrigation water, be careful about overhead application of soluble fertilizer. Especially if the finished product isn't washed, nutrient solution can leave off flavors and visible residue on the leaves where it dries. Fish emulsion should never be used for micro production for this reason.

The question I would ask myself if I were just getting started with microgreens would be: *How can I easily piggyback micro production onto my existing setup?* Because until I had determined the size of the market and how lucrative microgreens were going to be for me, I would want to get into micros as simply and cheaply as possible.

For instance, many growers simply put potting soil into flats and only ever use water for the full production cycle. The message here is that microgreens may be successfully produced under a range of conditions. Experiment with different production methods that are compatible with what you are currently growing. Growing with some fertility, in either the soil or water, will make microgreens grow faster and reach harvest more quickly. Slow-growing microgreens varieties, which may take up to a month to reach the harvest stage, especially

Figure 13.19. These trays were all seeded at the same time and covered with vermiculite. It's easy to see how the slowest-growing varieties would not make a good mix with the fastest ones.

Figure 13.20. The same trays a few days later. The fastest varieties are ready to harvest though the slowest have barely germinated.

benefit from some fertility over this relatively long growing time.

The most common way to grow microgreens in soil is to use standard 1020 open flats, and put an inch (2.5 cm) or so of potting soil in the bottom. The potting soil may provide enough fertility, or you can incorporate fertilizer into the potting soil before planting if you need more. Most growers either tamp the seed into the surface of the soil or sprinkle a thin layer of fine vermiculite over the seed to keep it moist through the germination process.

Trays may then be put in a germinator, on heat mats, or directly in the greenhouse if it's warm enough for germination. Trays going in a germinator or other dark area may be covered with plastic sheets or clear humidity domes. Trays germinated in a sunny greenhouse should be covered with white trays to keep them from getting too hot. The small amount of air in a tray under a dome can overheat very quickly.

Watering can be automated by using overhead sprinklers or overhead spray booms. No matter which watering method is used, one of the tricks to getting good microgreens production is to water the crop adequately without overwatering. Because the plants are small and not using much water, they are easy to overwater; it can smother and drown the roots if they are constantly sitting in water.

Seeding

Seeding method is a matter of preference in microgreens production. Some growers like to measure out a predetermined amount of seed, whereas others prefer to eyeball it. Some sprinkle seed on by hand, others like to use a mechanical seed tapper. It doesn't really matter as long as you get even distribution over your chosen medium.

Planting Density

The correct density for seeding microgreens varies by species and how large you intend to let the seedling get. Micros that are harvested at a younger stage – between cotyledon and one true leaf, for example – may be planted more heavily than micros that will be grown to the three-leaf stage. This will improve yield per unit of area, since the crop will be smaller and spend less time in production than a micro intended to be harvested between the two- and three-true-leaf stage.

Figure 13.21. A tray grown in soil ready for harvest. If you'd like a larger microgreen with more of a cotyledon, use a lower planting density. This tray is so dense, it's in danger of damping off if grown much longer.

These recommendations provide a starting point for seeding density. You will need to customize them to fit your system. If you seed too lightly, you will not maximize the potential of your growing area. Seed too densely, and your crop may become diseased due to poor airflow, or get leggy by harvesttime.

Labor

Microgreens do not require a lot of labor to maintain, since they are planted and harvested from the propagation medium a relatively short time later. Labor can be streamlined by automating the watering.

Cutting microgreens is commonly done with scissors, since the greens typically aren't strong enough to be cut with a knife. This time-consuming process can be sped up by using electric clippers of various types. The quickest way to harvest is to cut the substrate to size before planting and put the micros straight into containers for delivery to the customer when ready. If the product needs to be washed this adds time but overall the labor involved in producing microgreens is less than that for other protected crops.

Figure 13.22. This tray of basil micros is suffering from damping off due to being planted too densely and to poor air circulation.

Yield

Microgreens yields vary widely in terms of both total yield and how quickly they reach maturity. Both factors have to be taken into consideration, for if you need a certain amount of micros on a certain date, you need to seed accordingly.

Pests and Diseases

Microgreens are prone to fewer pests and diseases than most other crops because they are in production for such a short period of time. Most of the production problems affecting microgreens are symptomatic of systemwide issues, such as shore fly or fungus gnat infestation of the growing medium, or damping off. In such circumstances, deal with the conditions that lead to the problems. Fungus gnats or shore flies may be an indication of overwatering. Damping off may mean that plantings are too dense for proper airflow, or that there is a pathogen in the water.

Flavor

Flavor depends on the species grown. If the crop is stressed for any reason, it may become more strongly flavored. It's important to make sure that residues from the medium or fertilizer do not impart an off flavor to the micros. Otherwise, they grow for such a short period of time that there isn't as much opportunity to influence flavor as with a long-season crop.

Harvest

Microgreens can be harvested by cutting them just above the soil level and sold by weight. Many customers, such

Figure 13.23. Microgreens harvested using battery-powered grass clippers.

as chefs in busy restaurants, prefer the convenience of pre-cut microgreens for the ability to reach into a bag and sprinkle them on dishes that need a garnish, instead of having to cut them themselves.

Post-Harvest/Storage

For growers who sell cut microgreens, there are two schools of thought on washing them. Some growers avoid washing cut product because the greens are so fragile; the handling process may result in crushing or bruising the product. Also, much of the commercially available salad washing and drying equipment is prone to letting the microgreens out with the wash water, since the openings are designed for much larger leaves.

Some growers do like washing microgreens, however, especially if they use a substrate (like soil or vermiculite) that tends to get into the leaves; washing is the only effective way of getting little bits out of the leaves. Another advantage of washing, especially during hot weather, is that the greens are so small that putting them in cold water instantly cools them down. This can improve shelf life, especially when temperatures are warmer than ideal. Having a little bit of moisture on the leaves may improve shelf life as well, but as with full-sized greens, excessive moisture will cause the greens to rot more quickly.

Given the tender nature of the product, refrigeration is necessary to keep microgreens fresh. Temperatures from 41°F (5°C) down to just above freezing should work well to hold most species, with the exception of basil, which will blacken and spoil at temperatures below 50°F (10°C). Basil should be stored separately from other microgreens in order to avoid cold damage.

Marketing

Having a unique, distinctive microgreens mix will set you apart from other producers. Most microgreens are sold directly to restaurants, or to wholesalers who deal with restaurants, so getting to know chefs or buyers and acquainting them with your product may pay off. Some seed catalogs have nice pictures of what each individual species looks like and descriptions of the flavor. You can take a seed catalog or tablet computer in and show the chef all the many options for leaf shapes, colors, and flavors, and offer to make them a custom mix of their choosing — for a premium, of course. It's the kind of product that may demand a little extra work forming relationships with individual restaurants, but with their high price per unit and quick turnover, microgreens can be very profitable.

Figure 13.24. A plate in a restaurant garnished with microgreens.

Microgreens have a more limited market than the other crops in this book. They are not usually sold in farmers markets, CSAs, or grocery stores. For growers who have access to restaurants and high-end wholesale accounts, they can be rewarding. The typically high price per unit and rapid turnaround time mean that micros can generate very high amounts of income in a relatively small area.

APPENDIX A

Hydroponics

The term *hydroponics* refers to any system where the plant receives its nourishment from nutrients in the water. The water-based fertilizer is usually called nutrient solution. This is in contrast with soil-based growing, where the plants receive their nourishment from the soil.

In North America, hydroponic systems are almost always implemented in greenhouses rather than outdoors. Since there is no soil and not much medium to buffer the effects of rainfall, when an outdoor hydro system gets rained on it can dilute the nutrient solution feeding the crop. It's not unheard of to have a hydroponic system outside, but most of those are in dry areas where there isn't much rainfall. I became interested in hydroponics when I started visiting large commercial greenhouses and noticed that the vast majority used hydroponic systems. Even though I always planned on using soil myself, I started studying hydroponics in order to understand why it was used in the vast majority of large-scale commercial production.

Most of the commercial growers still using soil are certified organic. The United States is unusual in that hydroponic growers can, under current NOP (National Organic Program) rules, be certified organic. However, at the time of this writing there is controversy over whether hydroponics is compatible with the original intentions of organic certification, and there will probably be a ruling in the near future over whether or not hydroponics can continue to be certified. I agree with those who think that hydroponic systems should not continue to be certified as organic. I think this because the original intention of organics was to have biologically active soil, and to feed the plant by feeding the soil. Which, of course is impossible if you don't have soil. I think it's quite likely that if hydroponics is no longer certifiable, another label will emerge that will cover hydroponic growers who are as natural as possible.

Since this book is mostly about the plant management practices that happen above the ground, most of the information applies equally to soil and hydroponic growers. And there are already a number of books written about hydroponic production, so it's worth looking there if hydroponic growing is something you want to pursue.

Hydroponics Versus Soil

There are big differences between growing in the soil and in hydroponics, but either method can yield excellent results when managed properly. When many people use the term *high-tech* they mean hydroponic, but this is not always the case. I have seen some very high-tech greenhouses that are growing in soil. They can achieve yields comparable with high-tech hydroponic operations. As long as fertility is optimized, it doesn't matter what the medium is. This may be a more common concept in Europe, where there are more organic soil-based greenhouse operations, since growers can't be hydroponic and certified organic there.

There are advantages to growing hydroponically, which include being able to grow anywhere regardless of whether there is good soil, or any soil at all for that matter. Most of the largest greenhouses, as well as many small growers, use hydroponics for the same reason they put a greenhouse up in the first place: control. It's much easier to assess and monitor the fertility of a hydroponic crop, since you start with a blank slate with a hydroponic medium. You know what's available to the plant because you're giving it all of its nutrients. In soil

Hydroponics and Organics

There is some controversy over whether hydroponic growing methods are compatible with organic production. In Europe, organic production is not allowed to be hydroponic. There are growers in Europe who grow hydroponic organically and have their produce certified by North American certifying organizations and shipped to North America to be marketed as organic.

In North America, some certifiers will certify hydroponic crops organic as long as they meet all the other criteria for organic production. Other certifiers will not certify crops produced with hydroponics. The National Organic Program is in the process of trying to resolve the discrepancy.

In the past one of the barriers to organic hydroponic production was the lack of natural mediums and fertilizers compatible with hydroponic systems. This rules out rockwool and other synthetic mediums and most of the commonly used fertilizers for hydroponic production. The problem with using traditional organic fertilizers like fish emulsion in a hydroponic system is that they have a tendency to clog lines and grow algae. However, a lot of progress has been made on natural fertilizers that are compatible with hydroponic systems in the past few years. Likewise the popularity of cocoa coir as a medium for hydroponic growing has given growers a natural option for substrate.

growing, it's more difficult to assess nutrient availability, even if you know what's in the soil. Root diseases are also easier to deal with in hydroponics; given that blank slate, there's usually no disease carryover from the previous crop.

The advantages of hydroponics become apparent when contrasted with the difficulties posed by growing in the same soil year after year. Pests, diseases, fertility imbalances, and salts can build up in the soil over time. Hydroponics allows you to start fresh every year, with clean growing medium and nutrient solution.

As a soil grower, I've heard time and again from hydroponic growers and consultants that "sooner or later, you're going to get disease in there," which is probably true. It's almost inevitable that diseases will build up over time when you grow the same crop in the same soil every year. Crop rotations help. Movable structures help, though this is only an option for smaller hoophouses without a lot of infrastructure. Grafting helps. But realistically, there is a reason why protected culture is not famous for having great crop rotations.

Unlike field growers, who grow five or more families of crops that they can rotate through different parts of the field, greenhouse growers often grow only a few families of crops, based on what is most profitable. If tomatoes are the most profitable crop, many growers are unlikely to plant another crop just for the sake of rotation, especially if they only have one greenhouse or hoophouse. The other disadvantage that greenhouse growers face when it comes to crop rotations is that it's usually economically unfeasible to leave a greenhouse fallow for part or all of the growing season. Because field space is less expensive, many growers can afford to plant a cover crop on part of their field acreage every year to give it a rest from production. This isn't usually possible in a greenhouse; the cost of the space demands that it be planted with a cash crop every year.

The cost of growing something less valuable than the eight crops in this book is such that most growers can't afford to give up the revenue from their best crops. They likely can't weather a switch from growing tomatoes to growing carrots for a year in order to get out of the crop families that they normally grow. So instead of rotating the crop, hydroponics rotates the growing medium by tossing it out and starting with new medium every year.

Types of Hydroponic Systems

Between commercially manufactured and homemade systems, there are infinite ways to set up hydroponic growing systems, but they can be divided most broadly into whether the roots grow in a solid substance or in nutrient solution with no medium. There are exceptions, but commercially speaking, systems with mediums are most widely used with the fruiting crops in this book, while systems without mediums are most widely used for the leafy crops. So if you want to grow fruiting crops, consider container or slab culture.

The leafy crops are most often grown in liquid hydroponics. The frequent crop turns require frequent reuse or replacement of the medium, both of which are problematic from a labor, expense, and disease standpoint. So if you want to grow any of the greens crops, consider Nutrient Film Technique (NFT) or deep water culture (DWC) systems as described on page 230.

Hydroponic Systems with Mediums

In systems with mediums, the solid that the roots grow in is sometimes referred to as substrate. The medium takes the place of soil, though usually an inert material without any fertility is used to anchor the plant.

There are two basic types of substrate systems. Systems that let the nutrient solution drain out the bottom of the substrate and into the ground are known as drain-to-waste systems. The alternative is a recirculating system, where nutrient solution is collected when it drains out of the medium and pumped back to a nutrient reservoir to be used again. Recirculating systems are more complex because they have to be plumbed back to the nutrient reservoir, and also because the reused nutrient solution has to be carefully monitored and adjusted to make up for what the plants removed the first time through. Recirculating systems also need to be sterilized to keep diseases from building up in the nutrient solution.

Figure A.1. This young tomato crop is being grown hydroponically in a container designed for growing, called a Bato bucket.

Container Culture

Container culture refers to any method that uses loose substrate in a container. I've seen everything used from grow bags to buckets designed specifically for this purpose, as well as 5-gallon buckets, trash bags, or anything you can imagine that can contain a plant's roots. One variant of container culture involves making raised beds filled with substrate so you have one long bed instead of individual containers.

Slab Culture

Currently, most of the commercially grown hydroponic fruiting crops are grown in slab culture. Slab culture uses short slabs of medium, most commonly coir or rockwool, to anchor the plants' roots. Slabs are made in different sizes based on the crop, but they are usually 3 to 6 inches (8–15 cm) deep, 6 to 12 inches (15–30 cm) wide, and a few feet (1 m or so) long. Slabs are lined up end-to-end to form continuous beds onto which the crop is transplanted. If you're interested in slab culture, vendors can help you figure out the best size and medium for your crop. The medium is usually wrapped in plastic to contain the roots and moisture. They are popular with large operations because setup is fast: The slabs can be quickly lined up at the beginning of the season and removed at the end, making cleanup easy.

Hydroponic Systems Without Mediums

Hydroponic systems that do not use mediums to hold the roots are known as liquid hydroponic systems. There are two main systems for liquid hydroponics that are used commercially.

Nutrient Film Technique

Nutrient Film Technique (NFT) systems use long plastic channels to grow plants. There is no medium in the channel other than the plug the plants are started in. The term NFT comes from the fact that the roots are growing only in a thin film of nutrient solution in the bottom of the channel.

Figure A.2. An indoor crop of NFT lettuce. You can see the type of rockwool plugs the plants are growing in where there are missing plants.

The nutrient solution is constantly refreshed by flow from an emitter at the top of the channel. The channels are placed at a slight grade, so the nutrient solution flows down the channel over all the roots. At the bottom end of the channel, the nutrient solution is collected and recirculated. Usually the channels, also called gutters, are enclosed, with a hole where each plant goes. Open gutters can be used for the production of microgreens.

One of the advantages of NFT is that it does not require as much nutrient solution to grow the same area as deep water culture (see below). Also, NFT systems can be installed in otherwise open greenhouses without the need to build basins, so you may be able to be reconfigure them more easily. On the other hand, because there is a smaller amount of solution, it can go out of adjustment faster and needs more careful monitoring than DWC. Another drawback of NFT is that the solution needs constant circulation. If the power goes out, the crop will begin to suffer very quickly from the lack of circulation.

Deep Water Culture

Deep water culture (DWC) is a system where plants are grown floating in nutrient solution. Plants are grown in basins, anchored in rafts made out of inert foam that floats on top of the nutrient solution.

Deep water culture is sometimes preferred by growers who want to build systems themselves, because they can be built more simply than NFT. Simple DWC systems are basically a tub with some foam insulation floating on top with holes drilled to hold the plants. There are fancier commercial systems that recirculate and aerate the nutrient solution, and these types of improvements will increase yield.

One of the disadvantages of DWC is that you need more nutrient solution to grow the same amount of area as with NFT. Still, under some circumstances this can be an advantage, because a larger reservoir of nutrient solution will be more buffered against temperature changes in hot climates. And it will change composition more

Figure A.3. Butterhead lettuce growing in a DWC aquaponic system.

Figure A.4. Red leaf lettuce in DWC. Note the floating roots in the foreground.

gradually than a smaller reservoir, so you don't have to monitor the nutrient solution as carefully as with NFT. Overall a DWC system may be more forgiving than an NFT system.

Aquaponics

Aquaponics is a variation on hydroponics where the waste from aquaculture provides the main or only fertility in a hydroponic system. There is currently a lot of interest in aquaponics as a sustainable option since the fish (or other aquatic creatures) provide the fertility, and then become food themselves. This is a relatively new idea for commercial growing, though there are commercial aquaponic operations. One of the dynamics of aquaponics is that growers tend to break even on the fish, so aquaponics systems have to be evaluated on the merits of providing fertilizer, and the profit must be made on the produce.

APPENDIX B

Pests and Diseases

This appendix represents a very general overview of the pests and diseases that are common to, if not unique to, protected culture. It's not a catalog of every pest and disease that can affect every crop in this book. That is a whole book in itself, my favorite being *Knowing and Recognizing: The Biology of Glasshouse Pests and Their Natural Enemies* by M. H. Malais and W. J. Ravensberg. Not only is it the most complete book on the subject that I am aware of, but it has a large format and lots of great pictures. It's basically an incredibly detailed coffee table book on biocontrols. Its large size is really handy when it comes to getting an up-close view of organisms that are measured in fractions of a millimeter. And it goes into depth even on some of the less common pests.

A lot of progress is currently being made in the realm of biocontrols. New beneficial insects are being discovered all the time. So it's important not only to rely on printed materials but also to look online and get in touch with biocontrol suppliers to find out what the best option is for dealing with a particular pest. The foreword to *Knowing and Recognizing* is titled "Biological Control Does Not Stand Still," and there may well be new biological solutions to your pest problems by the time you read this. But the book will remain a good overview of pests and beneficials, and it will always look great on your coffee table.

Geography is a major factor affecting pest and disease problems in protected culture. What pests and diseases are a threat to your protected growing is greatly determined by what's active in your area and time of year. For example, some pests and beneficials are more active in hotter or cooler weather, so your solution has to be tailored not only to the problem but to the environmental conditions as well.

For the most relevant information on what pests are likely to be a problem for you and how to deal with them, a biological control company that services your area is your best bet. For disease control options beyond those suggested in this book, contact a greenhouse supply company. Protected culture does not so much eliminate pests and diseases as make some of the diseases of the field less likely and others more likely. The specifics of how they impact each crop are contained in the individual crop chapters.

Integrated Pest Management and Biological Control

Even for growers who are not certified organic, the trend in greenhouse management is toward using prevention and biological (nonchemical) means of controlling pests. This is known as Integrated Pest Management, or IPM, since other methods are integrated into controlling pests besides just spraying. Reasons for this trend include the fact that it's more difficult to spray in a confined area than it is in the field. Also, most pesticides have a reentry interval wherein workers are not supposed to be in the crop after spraying. Which doesn't work for greenhouse crops that need to be maintained on a daily basis.

Marketing produce as "pesticide-free" has also become a way for some greenhouse growers to set their produce apart from conventional. Very small pests like thrips can be difficult to kill with contact insecticides, since they feed deep in plant parts where they are hard to reach with spray. Pests with short life cycles like whiteflies are notoriously quick to become resistant to pesticides, even for growers who would love to spray them.

A final reason why so many growers are switching entirely to biological controls is the incompatibility of beneficial insects with spraying. If you're using bumblebees or other beneficial insects, it's difficult not to kill them when you spray insecticides. So for all these reasons many growers, regardless of whether they are organic, are switching to biological control. If you plan to use them, have a plan for beneficial insects based on the pests you anticipate. In many cases this includes releasing beneficials in your structure *before* the pests arrive, so they are waiting and hungry when pests show up. Being proactive to keep problems under control, rather than following the reactive mind-set of spraying, is what works best for biocontrol.

If you don't know what to expect, among your best allies are biological control companies. Most of the ones that are oriented toward commercial growers have programs where you can contract for a whole year's supply of biologicals at a fixed rate. This will include beginning to deploy biologicals shortly after transplanting, so that when pests show up there are already active predators. One advantage of a contract is that it will give you a plan for what your biological control strategy should look like from the beginning of the season. Which biologicals you use is highly dependent on the pests you are trying to control, anticipated temperatures, season of the year, and the crops you are growing. A complicating factor is that new biological controls are being discovered all the time. Yet another advantage of getting a contract with a biological control company is that your plan will be devised by a pest control specialist with up-to-date knowledge of the field.

Pest Prevention and Crop Hygiene

The best way to keep diseases from becoming established in your greenhouse is to keep them out in the first place. The beginning of each new season gives you an opportunity to make sure your structure starts out clean. Do a thorough cleanout between crops, removing all plant debris and cleaning any parts of the structure that might harbor debris, pests, or pathogens from the last crop.

Ideally the structure will sit empty for at least a week or two if not more between crops to exhaust any pests that might still be present—though this isn't always possible. Crops remaining in a structure are sometimes referred to as a green bridge, carrying pests and diseases over from one crop to the next. Next, think about all the ways that pathogens can get into your growing area, and eliminate any that you can. There may be some that are endemic or carried on the wind and difficult to exclude. But the first line of defense should be protocols to block any easy ways that pathogens might make it in. Especially if you also have field operations on your farm, staff should never go from working in the field to working in a greenhouse or hoophouse. That's a great way for diseases that start in the field to make it into your protected growing area.

In a controlled environment, excluding pests should be considered to reduce pest pressure. Even a simple practice like keeping the sides rolled down at night excludes the moths that lay tomato hornworm eggs, as well as other night-flying pests. Good hygiene practices can help keep pests out of the greenhouse as well. One good practice is never to bring houseplants or anything besides the production crop into the structure. Just because they aren't an agricultural crop doesn't meant they aren't harboring pests that may become established in your crop.

In areas where flying pests are a major problem, screening all greenhouse openings can significantly reduce pressure. Just make sure you use screen that has mesh smaller than the organism you're seeking to exclude. The downside to screens is that no matter how large the mesh, screens will slow down airflow, which can lead to a hotter greenhouse. When screens are used, it may be necessary to increase the surface area of screening to make up for the loss in airflow by building boxes over openings with as much as five times as much surface area as the opening being screened.

Some growers try to minimize the amount of vegetation around their greenhouses and hoophouses, thinking it harbors pests that may come into the greenhouse. Other growers have been successful with the opposite

Trading Field Pathogens for Protected Culture Pathogens (Happily)

My own tomato and cucumber production offers a real-world example of how protected growing may result in a more manageable disease load than field growing.

As I mentioned in the introduction, we moved all of our large-fruited tomato production indoors after experiencing roughly a 50 percent rate of second-quality fruit in the field. This lifted our packout rate into the 90s. As soon as we moved the crop indoors, the incidence of septoria leaf spot, early blight, and late blight went down almost to zero, since these diseases are promoted by rain. Though not a disease, splitting went down to almost none as soon as we put the plants under cover.

On the other hand, within a couple of seasons we started getting leaf mold every year. This was easily controlled with resistant varieties. On the pest side, we traded hornworms in the field for occasional whiteflies in our hoophouses. On our cucumbers, we still get cucumber beetles, but at least we can exclude most of them with insect netting. So some of the pests are the same as in the field, some are different, and either way you have more control options in protected culture.

approach: making plantings designed to attract and feed beneficial insects, a technique known as farmscaping. It may be possible to increase the number of beneficial insects around your structure by releasing desired species when the plants they like are in bloom. For example, many species in the umbel family, and other plants with small flowers that bloom over an extended period, can help feed beneficials. Many beneficial insects feed on flowers along with pests. Of course there can't be any screens over openings if you want the beneficials to come in from the outside.

Pests

Most of the pests affecting greenhouse crops are insect, moth, and butterfly species. I will touch briefly on each species type that is most problematic in greenhouse production and give a basic overview of the best strategies for dealing with them, but I'm not going to make recommendations for specific chemical and biological products, since the names change so quickly. Contact the appropriate supplier for the most up-to-date information. There are biological controls for all these pests.

Aphids

Many different species of aphids may feed on greenhouse crops. In the process of piercing plant tissues and sucking out juices, they damage and deform the plant. Aphids can vector diseases by feeding on infected plants and moving to healthy ones. Populations can explode very quickly. In addition to feeding damage, they also secrete honeydew and leave cast-off exoskeletons, which can make a mess of produce and plants.

Thrips

Many different species of these tiny insects can be a problem in greenhouses. In protected environments without predators, populations can get out of control quickly. They suck the juices out of plants, causing damage and potentially transmitting pathogens.

Whiteflies

Two species of these minuscule insects most commonly become problems in greenhouses: the greenhouse whitefly (*Trialeurodes vaporariorum*) and the silverleaf or sweet potato whitefly (*Bemisia tabaci*). If you see very tiny insects that are almost entirely white feeding on your plants, they are probably one of the whiteflies.

Not only do whiteflies suck juices out of plants and transmit diseases, they also excrete large amounts of honeydew. Honeydew can be as much or more of a problem as the feeding damage itself, because it can grow mold and necessitate cleaning of greenhouse crops that otherwise don't need to be cleaned.

Tomato Hornworms

These large and voracious caterpillars are the larval stage of two species of night-flying moths. They are mostly a pest of tomatoes, but will occasionally be a nuisance on other solanaceous crops. Screening or rolling down sides and closing vents at night will exclude them.

Mites

A number of different species of these small, spider-like creatures can be pests of protected culture crops. In addition to feeding on leaves, which can cause leaf death, they may also damage fruit and spin webs, which can ruin the appearance of crops. There are lots of predators that feed on mites.

Butterflies and Moths

A number of butterfly and moth larvae may cause damage to leafy crops, with brassicas in particular being prone to caterpillar damage. Caterpillars consume a lot of leaf area as they grow. Some growers have a zero-tolerance policy on selling produce with caterpillar damage, since if you see the evidence there is likely a live caterpillar and frass (poop) on the plant as well.

Diseases and Disorders

Just as with pests, protected growing offers more options for controlling diseases and disorders than does open-field growing thanks to the greater amount of control.

Blossom End Rot and Tipburn

Blossom end rot (BER) is a common problem in tomatoes and peppers, less so in eggplant; and the same disorder is responsible for tipburn in lettuce. It's not a disease but rather an abiotic condition caused by inadequate levels of calcium in rapidly developing plant parts. Both disorders are symptoms of calcium deficiency in the fruit, or leaves, in the case of lettuce.

In tomatoes and eggplant, the disorder may present as small brown spots or leathery patches on the blossom end or sides of fruit. In peppers, blossom end rot is frequently confused with sunscald. You can tell the two apart because damage from blossom end rot appears tan and is usually lower down on the fruit; sunscalded flesh looks white and is usually higher up, near the shoulders where the fruit gets hit with direct sunlight. Tissue affected by blossom end rot is susceptible to secondary infections, most notably botrytis, which can give the affected area a dark gray, fuzzy appearance (see figures B.1–B.5).

Figure B.1. Controlling the environment and crop care can be used to solve many of the problems in protected culture. A good example is this immature tomato, whose flower is infected, with botrytis. Botrytis (see more in chapter 6) is a pathogen that affects almost all crops. At first this would seem to be a disease issue. But since botrytis spores are almost everywhere, there isn't much chance of excluding it from your growing area. Still, you can control the pathogen through management of temperature and humidity. Botrytis tends not to develop under warm temperatures with low humidity. And blossoms of all the fruiting crops tend to stick to the fruit in high humidity with poor air movement.

Figure B.2. If the blossoms do fall off but the humidity is too high and temperature too low to prevent botrytis, they may start an infection on the leaves like this.

Figure B.3. If not dealt with promptly, the infection may take over the leaf, like this. The only way to deal with it now is to prune the leaf off, cutting it flush with the main stem so as not to leave a stub that could become infected.

Figure B.4. If diseased fruit isn't removed, it can fall off and infect other plant parts. Plant parts infected with botrytis are the zombies of the greenhouse world—they can spread the disease to other plant parts as they fall through the canopy. You can see why the common name for botrytis is gray mold.

Figure B.5. This is a spectacular display of what happens when botrytis is left to develop unchecked. The gray mass at the front is the tomato that started the infection. These photos are a good example of how timely crop care could have prevented more problems later.

In lettuce, tipburn usually appears on the youngest, most rapidly growing leaves at the center of the head. It can appear as black spots or brown discolored areas along the margins of the leaves, which may give the leaves a singed look.

Cause

BER and tipburn are classic signs of calcium deficiency, one cause being an inadequate amount of available calcium. The disorders are tricky, because there are many reasons why a plant might exhibit symptoms even though there is enough calcium in the growing medium.

Large fluctuations in watering, including too much water or excess drying, can interfere with the uptake of calcium and cause BER. So if there has been flooding or an interruption of water supply, fruits that were developing at the time of the disruption may show symptoms. Another common cause of BER is excessively vegetative growth, due to either the stage of plant growth or excess nitrogen fertilization. Plants suffering from BER for this reason may even show adequate calcium levels in tissue testing. During excessively vegetative growth, most of the calcium may be drawn to the rampantly growing leaves, starving the fruit.

Tipburn is most common when high light levels are fueling rapid growth, or when high humidity or poor ventilation causes low levels of transpiration. As the plants respire, especially in stagnant air, humidity can build up around the leaves. This is the most pronounced in the depression in the middle of head lettuce where new leaves are forming. If humidity builds up enough, respiration stops because the air can't accept any more moisture from the plant. Respiration is how leaves bring new fluids in with nutrients, so they will become starved if respiration stops.

Prevention

There is no cure to reverse the effects of BER once it shows up in fruit or leaves, so efforts to deal with the disorder focus on prevention. Fruits that show symptoms should be removed. They will not recover and become marketable. Likewise, lettuce with tipburn should be removed to make space for growing something marketable.

Making sure there is enough calcium in the growing medium is the first step to preventing the disorder. Ensuring even watering, so the plants always have enough water and not too much, is also a good management technique to keep BER in check. Avoiding nitrogen overfertilization is one way to help keep plants from becoming too vegetative. Techniques to help balance the plants if they become too vegetative are described in chapter 7.

BER can be common in young tomato plants, since they tend to be vegetative early in the growth cycle. Calcium is drawn to the plant parts that are respiring the most, so if there's a lot of leaf area and very little fruit load, the leaves may end up hogging the calcium. Crops tend to balance out over time as they become more mature and less prone to BER, especially if you can help steer them in a more generative direction when they are too vegetative.

To prevent tipburn, it may be necessary to shade the structure in summer when growth is too fast. This will also help keep the greenhouse cooler to help with bolting. If high humidity or poor airflow is contributing to tipburn, horizontal airflow and even vertical airflow fans can help reduce stagnant air.

Russeting

Russeting is the development of micro cracks on the skin of the tomato fruit. It is known as cuticle cracking, or rain check when it occurs in the field. It's differentiated from other cracking disorders by the size of the cracks. Instead of large radial or concentric cracks caused by too much moisture, russeting produces smaller, interlocking cracks. These can range in size from almost invisible hairlines to plainly visible fissures. Russeting is a problem not only because it's unsightly, but also because it decreases the shelf life of the fruit. Moisture escapes through the cracks, causing the fruit to dehydrate more quickly, and pathogens can enter, causing premature rotting.

Russeting is a complex disorder that has a number of contributing causes. One factor is genetic, and some tomato varieties are more prone to it than others. If you have problems with russeting, trying other varieties may help. Another contributing factor is a widely fluctuating moisture level as the fruit approaches ripeness. When there is excess water the plant puts more into the fruit, and when there isn't enough it takes water out. Fluctuations cause imperceptible swelling and contractions in the fruit, causing it to lose elasticity. The more this process occurs close to ripeness, the more likely it is that hairline cracks will occur on the fruit.

A very large day/nighttime differential, excessively high daytime temperatures, high relative humidity, and condensation on the fruit are all factors predisposing plants to russeting. Sunlight falling directly on the fruit may make the problem worse, if the canopy is very sparse or leaves have been removed from over the tomatoes. Adhering to the temperature recommendations in this book will go a long way toward minimizing the disorder. However, if you're growing in an unheated hoophouse or another situation where ideal conditions can't be maintained, this may help explain the appearance of russeting.

Russeting can be common in unheated hoophouses in the fall, when there's a huge day/night temperature differential. The other factor predisposing fall hoophouse tomatoes to russeting is the condensation caused when the sun comes up and warms the air. The plants and especially the fruits stay cold longer than the air, and the moisture from the night's transpiration condenses on the plants and the fruit, making them wet even in the absence of rain.

Downy and Powdery Mildew

Downy and powdery mildew are fungal diseases that affect a wide range of greenhouse crops. There are different species of the pathogen that cause similar symptoms on different crops; the species of powdery mildew that affects lettuce, for example, is different from the species that affects tomatoes. Both pathogens colonize leaf surfaces and reduce leaves' ability to photosynthesize.

Though use of resistant varieties is the best strategy for dealing with these pathogens, greenhouse environmental control can help with disease management in nonresistant varieties. It's important to know the preferred conditions of each pathogen so you can create conditions that aren't conducive to its development. In some areas, downy and powdery mildew are year-round problems. In areas where the pathogens don't overwinter, spores can blow into a completely clean structure and cause an outbreak.

Fusarium, Verticillium, and Other Soil Diseases

One of the reasons hydroponics is so popular is that these diseases are not usually found in soilless systems. Soil growers should prevent transfer of potentially infected soil from field production areas and use resistant varieties where possible. If nonresistant varieties must be grown in infected soil, many species have resistant rootstocks available; see chapter 8.

Viruses

There are many viruses to which protected culture crops are susceptible. They can be a real problem, because an extended season provides a long time for a virus to spread throughout the crop. This may occur through vectors like whiteflies, or workers using the same tool on one plant after another.

Varieties have been developed that are resistant to many viruses. If you're not using resistant varieties, make sure plants enter the production area healthy by using clean seed and keeping a clean propagation area. Screen seedlings for symptoms of viruses, since it only takes one plant to start an outbreak. Once plants are infected, they are not curable.

APPENDIX C

Tools and Supplies

There are a number of specialized tools for use in protected culture. What follows here isn't a comprehensive list, but it does include the most commonly used tools. In most cases it's best not to store tools and supplies in the greenhouse when they're not in use, as the humidity will accelerate rusting of metals. The high level of ultraviolet light also degrades many plastics. If you have any fungal pathogens that are sporulating, they will settle on everything, including tools. You don't want your tools covered in spores when you go to use them; they have the potential to spread the pathogen to every plant you work on.

Trellis Clips

Trellis clips are used to secure vining crops to trellis twine and keep them upright as they grow. As the plant gets taller, clips are added every foot (30 cm) or so before the plant flops away from the string. This is an alternative to twisting the string around the vine as it grows (see "Trellising" in chapter 6).

Clips have a hinge on one side and a latch opposite, and are usually about 0.75 inch (2 cm) in diameter. The hinge is made so that when it's closed on a piece of twine, it will grab the twine and hold the clip securely in place. The latch can be reopened and repositioned if it needs to be moved.

Clips are available in regular and compostable plastic. Though biodegradable clips cost a little more, they are worth it because they speed end-of-season cleanup. If you use biodegradable twine, you can simply cut the vines down and compost them – twine, clips, and all – instead of having to remove the vines from the noncompostable parts.

We've developed a signaling system on my farm using these clips. When workers spot a potential problem or question with a plant, they put a clip on the string 1 or 2 feet (30–60 cm) above the plant. This makes it easier to locate plants that need special attention later. Bright flagging tape would be more eye catching than a clip, but clips are what we have when we're doing plant maintenance. A common reason we flag a plant is if it has a broken or damaged head, reminding the next person who works on it to leave some suckers to develop a new head.

We also put a clip above a plant that's showing signs of disease, as a signal not to work on that plant and spread disease to the other plants. When I'm not around, workers flag plants that have anything they don't know how to deal with, so later we can quickly identify the problems and figure out what to do about them.

Clippers

For most greenhouse uses where cutting is necessary, standard farm or garden scissors or bypass-type clippers can be used. Anvil pruners are for deadwood and should not be used on live plants, as they may leave a ragged cut. It's best if whatever you're using is all metal and plastic with a smooth surface; tools with a lot of texture can be hard to sanitize. Blunt-tipped clippers will minimize the amount of accidental puncturing of stems and fruit that can occur with pointy scissors.

I particularly like the Saboten line of harvest scissors, which are small enough to fit in the palm of your hand. Given their compact size, it's easier to make cuts in and around tightly spaced plants. They also have some ingenious features – you can wear the handle like a ring, for

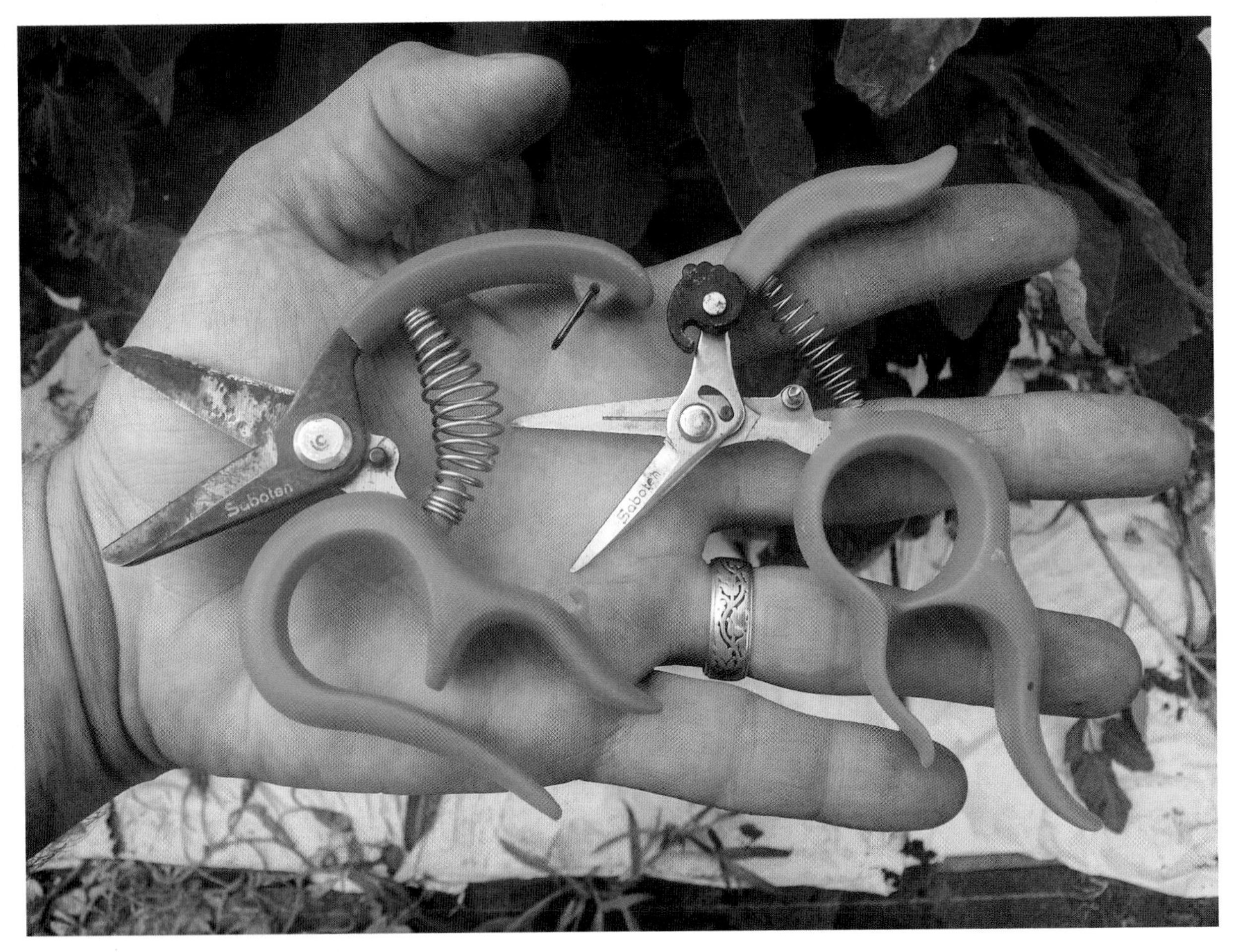

Figure C.1. I like the clippers on the right for pruning, because they can be worn on a finger then shifted out of the way when you need to use your hands; the safety can be operated by your thumb.

Figure C.2. Here's the tool rotated out of the way so you can use your hands.

instance, so you can flip the tool out of the way and work two-handed without putting it down. Some models have a safety that keeps them closed when not in use; this is operated by the thumb of your working hand, so you can use them completely one-handed.

Pruning Knives

A lot of growers prefer to prune leaves off with a knife, because it leaves a cleaner cut that's more flush to the stem than most pruners. There are specialized short, blunt-tipped, single-sided pruning knives made specifically for pruning jobs such as de-leafing vining crops. The blade is designed to minimize unintentional cuts and punctures when you're working fast in tight spaces. X-Acto blades in holders and razor blades also work well.

Ring Knives

Ring knives are small, curved knives attached to a loop that can be worn on a finger. They're designed to speed up jobs that require many small cuts and both hands. The stems of most other fruiting crops are too thick, but using a ring knife to harvest cucumbers may speed up the process.

Fingers

I do any job that I possibly can with my bare hands. For example, I always try to snap suckers off with my hands rather than cutting them – especially when I can do it in the morning when the plants are turgid. They usually pop off pretty cleanly, and I don't have to fool with a tool. Another advantage of using your fingers is that they are less likely to transmit diseases from plant to plant, unlike a blade that comes in contact with the vascular system of each plant.

Tomahooks/Spools

There are several methods used to lower and lean vining crops, enabling you to grow a vine longer than the greenhouse is tall (see chapter 6). The idea behind all the methods is that a spool of twine is attached to the overhead wire. A plant is trellised up the string. When the plant reaches the wire, some of the twine can be spooled out, lowering the plant and giving it more room to grow.

The tomahook is used by winding string around a wire hanger. When you want to lower the plants, the hanger must be removed by hand and twine manually spooled out. For tools like Rollerhooks or RollerPlasts, a circular spool releases twine when lowering is desired. All these methods were designed to hook onto overhead wires and do not work very well if you are using pipes for support. Pipes are too large for any of these tools to hook onto. If you must use pipes with any of these methods, add wire loops to the pipes at every point where you need to anchor a plant.

Truss Supports

We don't always use them, but when we notice that peduncles (the stems that hold the fruit) are too thin, we use truss supports. When a tomato crop is young or anytime the plants are very vegetative, they may produce weak peduncles. We have had beefsteak tomatoes with such weak stems that they ripped off of the plant.

There are two main forms of truss supports: the arched truss support and the J-hook. Arched truss supports look like a miniature plastic bridge that is snapped onto the truss where it comes off the main vine. The arched type is usually used on smaller tomatoes from grapes and cherries up to cluster tomatoes. Because the clusters on these smaller types of tomatoes aren't as heavy as on

Figure C.3. A fruit cluster that ripped off under its own weight.

Figure C.4. A truss hook properly applied on some beefsteak tomatoes. It needs to be applied below at least the first fruit so it doesn't slide down the stem as the fruit develops.

Figure C.5. A truss support being used on a cluster of tomatoes on the vine.

beefsteaks, all you need to keep them from kinking is a little support so they stay arched.

Large-fruited tomatoes can be so heavy that they need something more substantial to support them. This is done with J-hooks, which have a hook on one end and a clip for attaching to the trellis twine on the other. They are used by slipping the hook under the stem of the first or second fruit closest to the plant, and clipping the other end to the trellis string. This needs to be done when the fruit are small, before they have begun to exert much downward force on the peduncle.

Ergonomic Tools

I like using carpenters' nail pouches, which you can get at the hardware store, for holding small tools while I work, keeping my hands free. Just tie them around your waist like a belt and they hold enough clips to keep you busy for a long time. Since they're made of canvas, they can be laundered and sanitized.

For jobs that only require a little extra height, I like kick stools because I can push one down the aisle with my foot as I work and never have to stop what I'm doing to move it. The one I have is about 16 inches (41 cm) tall, made of plastic, and very light. It's the same thing you can find in libraries everywhere for grabbing books off the top shelf. Spring-loaded wheels in the base push the whole structure up 1 or 2 inches (2.5–5 cm) off the ground when there's no weight on it. As soon as you put weight on it, the springs compress and the base makes contact with the ground, stabilizing the whole thing. Stepladders can be used for taller jobs. The top wires in my hoophouses are 9 feet (3 m) off the ground, so lowering and leaning takes a little more height than a kick stool.

The ultimate in efficiency is a self-propelled scissors lift that runs on a pipe rail heating system. Acceleration is controlled by foot pedal. You can work at any height thanks to the scissors lift. They are not cheap, but in a big greenhouse they pay for themselves pretty quickly.

For close-to-the-ground work, one thing we have come to appreciate is planting carts that are designed to provide a mobile seat and toolbox in the garden. They're great for sitting on to make low-down jobs like de-leafing and harvesting less backbreaking. You can also bungee a bulb crate or box to the top and push it down the aisle with a foot, so you don't have to pick up boxes of produce as you harvest.

Duct Tape

In addition to everything else it can fix, duct tape can patch a broken tomato vine. When I have accidentally snapped them during lowering, I have had almost 100 percent success repairing vines with duct tape. This won't work if there's no attachment left, but usually when tomato vines snap the core breaks and a small part of the outer layer remains. This is all you need to potentially heal the plant.

As quickly as possible after a tomato plant is snapped, line the vine back up as it was before it broke and wrap the stem below the break with tape. Then wrap in a spiral up from the encircled area to contain the break entirely in duct tape, continuing on a few inches (cm) above the break so the "cast" is supported on both sides by healthy vine and duct tape. It may help to wrap a J-hook or other small piece of clean material under the tape along the break to act as a splint.

This has worked for me even on hot days when I was sure a broken-stemmed plant was a goner. One time when I couldn't find the duct tape I used electrical tape successfully – but duct tape definitely works better with its greater width and stickiness.

Sanitizers

Tools that are used on multiple plants should be sanitized with a frequency corresponding to the level of disease pressure. For example, if you're working in a fairly clean greenhouse with no active pathogens beyond a little botrytis, sanitize when you start using a tool, when you're done with it, and after using it on an infected plant. Growers with very high disease pressure may sanitize between each plant. That sounds extreme, but when an aggressive pathogen is getting out of control, it's better than spreading the disease to every plant. There are clippers available that feature a tube attached to a backpack reservoir of disinfectant. Every squeeze of the clippers squirts disinfectant onto the blades to eliminate the extra step of disinfecting.

Depending on what pathogen you're trying to control and what's permitted, you can use rubbing alcohol, bleach, or a disinfectant designed specifically to kill viruses, like Virkon. Before choosing a disinfectant, make sure it's effective against the pathogen of concern and legal for agricultural use.

Further Reading

Seeds, Tools, and Supplies

Hort Americas. www.hortamericas.com

Hydro-Gardens. www.hydro-gardens.com

Johnny's Selected Seeds. www.johnnyseeds.com

Paramount Seeds: Non-GMO seeds for commercial growers. www.paramountseeds.com

Educational Programs

The University of Arizona. The University of Arizona offers some excellent programs on greenhouse growing, both for undergraduates and some that the public can attend.

Books

Coleman, Eliot. *The New Organic Grower: A Master's Manual of Tools and Techniques for the Home and Market Gardener.* 2nd ed. Chelsea Green Publishing, 1995.

Coleman, Eliot. *The Winter Harvest Handbook: Year Round Vegetable Production Using Deep-Organic Techniques and Unheated Greenhouses.* Chelsea Green Publishing, 2009.

These two books by Eliot Coleman helped build the foundation of a lot of the techniques in this book.

Estabrook, Barry. *Tomatoland: How Modern Industrial Agriculture Destroyed Our Most Alluring Fruit.* Andrews McMeel Publishing, 2012.

An eye-opening depiction of big commercial field tomato production in Florida.

Hartman, Ben. *The Lean Farm: How to Minimize Waste, Increase Efficiency, and Maximize Value and Profits with Less Work.* Chelsea Green Publishing, 2015.

Not about greenhouse growing per se, but excellent for evaluating the efficiency of any farm or greenhouse business.

Malais, M.H. and W.J. Ravensberg. *Knowing and Recognizing: The Biology of Glasshouse Pests and Their Natural Enemies.* Reed Business Information, 2003.

The best book on the biocontrol of greenhouse pests.

OMAFRA Staff. *Publication 836: Growing Greenhouse Vegetables in Ontario.* Ontario Ministry of Agriculture, Food and Rural Affairs, 2010.

A great reference, very focused on hydroponics.

Wiswall, Richard. *The Organic Farmer's Business Handbook: A Complete Guide to Managing Finances, Crops, and Staff – and Making a Profit.* Chelsea Green Publishing, 2009.

Not a greenhouse book, but with the emphasis on record keeping, precision and efficiency, this book is a great help in these areas.

Wittwer, S.H. and S. Honma. *Greenhouse Tomatoes, Lettuce, and Cucumbers.* Michigan State University Press, 1979.

Out of print and out of date, though still available second hand and an interesting look at greenhouse growing from just before the age of hydroponics. Most of the resources since this time have focused on hydroponics.

Index

About the Author

ANN MEFFERD

Andrew Mefferd spent seven years in the research department at Johnny's Selected Seeds, traveling around the world to consult with researchers and farmers on the best practices in greenhouse growing. He put what he learned to use on his own farm in Maine. He is now the editor and publisher of *Growing for Market* magazine. Previously, he worked on farms in six states across the United States before starting his own farm. Andrew also works as a consultant on the topics covered in this book. For more about the magazine, please visit www.growingformarket.com. For more information on consulting, see www.andrewmefferd.com.